YOUR TELEPHONE
Operation, Selection,
and Installation

Martin Clifford was born in Harlem in Manhattan and is a graduate of the City University of New York. Subsequently he worked as an engineer for Sperry Gyroscope Co. with emphasis on the areas of voltage regulated power supplies, radar, and automatic direction finders for aircraft. He also acquired various licenses — radio amateur (ex W2CDV), first-class radiotelephone (commercial), and a license to teach vocational subjects from the University of the State of New York.

When the demand for engineers grew slack he became a teacher in private vocational schools and specialized in electronic communications. During this time he became interested in writing and gradually became a full-time free lance writer and has had numerous articles published in newspapers, trade, and consumer magazines. His articles have been syndicated in as many as 200 college newspapers.

Most of his writing efforts are devoted to books. He is the author of many books as well as a number of booklets on various topics in electronics, and several correspondence courses in electronics, radio, television, and drafting. He is the author of *The Complete Guide to Car Audio*, *The Complete Guide to High Fidelity*, and *The Complete Guide to Security*.

YOUR
TELEPHONE
Operation,
Selection,
and Installation

by MARTIN CLIFFORD

Howard W. Sams & Co., Inc.
4300 WEST 62ND ST. INDIANAPOLIS, INDIANA 46268 USA

FIRST EDITION
FIRST PRINTING — 1983

International Standard Book Number: 0-672-22065-2
Library of Congress Catalog Card Number: 83-50377

Edited by: *Welborn Associates*
Illustrated by: *R.E. Lund*

Printed in the United States of America.

PREFACE

In November 1975 the Federal Communications Commission (FCC) issued an order which released the control of the manufacture and rental or sales of telephones from the grip of the telephone utility. This ruling was challenged in the courts, but by October 1977 the decision of the FCC came into effect. Briefly, the FCC decision made it legal for consumers to buy and install their own telephones, and further, permitted consumers to wire their own homes or offices.

There is no question that the Bell System produced a method of telephonic communications that is, in the words of Ian M. Ross, President of Bell Laboratories: "Possibly the best example of high reliability is the U.S. telephone system, the world's largest computer. It must operate with virtually no downtime because 200 million people expect service when they need it, at any time of the day or night."

However, the disassociation of the telephone utility from its exclusive right to the telephone business has brought about a number of benefits. Your monthly telephone bill can be reduced because it is cheaper to buy and install your own telephones than to rent them. It has stimulated the marriage of the telephone and electronics, and so the telephone is well on its way to being more than just a basic communications device. Further, telephones are changing outwardly, not just inwardly. There are now available basic phones, premium, decorator, modern, and advanced phones. Basic and premium are similar to those rented from the telephone company. Decorator phones include nostalgic styles such as European cradle phones and early American candlestick phones. Telephones are even being made to appeal to children.

Preface

Today, some 140 million telephones are in use in homes in the U.S. Marketing estimates indicate that in five years private ownership will multiply to 95 percent of all telephones in use. Hence, the current potential for growth in the telephone business is virtually unparalleled by any other American industry.

According to a survey made in late 1981 by United Media Enterprises to 1000 people selected at random throughout the U.S., talking on the phone to relatives or friends was the most popular in-home spare-time activity, following television watching, reading a newspaper, and listening to music.

The purpose of this book is to let you know more about telephones, how they are connected, how to economize on your telephone bill, what telephone features to look for, how telephones are used in mobile systems, the different types of telephones, how to install your own telephones, telephone answering machines, how to use the telephone to increase your personal security, and some facts about analog and digital technology as these apply to telephones.

The telephone is changing and doing so rapidly. It can do more, much more than it ever has been able to do in the past. It has become more than just a calling and answering device. The purpose of this book, then, is to acquaint you with the telephone's new look, both technically and decoratively.

MARTIN CLIFFORD

ACKNOWLEDGMENTS

A number of manufacturers of telephones, add-on devices and accessories were highly cooperative in supplying technical information, and not only manufacturers but suppliers as well were of considerable assistance. I was fortunate in being able to meet and talk with many of them, and to examine and discuss their product lines at the Consumer Electronics Show.

I would especially like to thank American Telecommunications Corp., the Bell Systems, and GC Electronics.

The address of every manufacturer whose products are illustrated are listed in the back of this book. These are:

All-Channel Products (Electro Audio Dynamics, Inc.)

Almicro Electronics, Inc.

Anova Electronics

Autumn Company

Bell Systems

CCS Communication Control, Inc.

Chrono-Art, Inc.

Code-A-Phone (Ford Industries)

Comdial Corp. (American Telecommunications Corp.)

Comvu Corp.

Acknowledgments

Contec Electronics, Inc.

Cose Technology Corp.

Dictograph Corp.

Dur-O-Peg

Dynascan Corp.

Electra Co. (Masco Corp. of Indiana)

Fanon-Courier Corp.

Fone Booth

GC Electronics

GCS Electronics, Inc.

General Electric Co.

GTE, Consumer Sales Div.

Harris Corp.

International Mobile Machines Corp.

Maxon Electronics Inc.

Midland International Corp.

Mitel Corp.

Mura Corp.

New Horizons

Nichco Inc.

Nortronics Co., Inc.

Panasonic (Consumer and Home Electronics Groups)

Pathcom, Inc.

Phonesitter Corp.

Recoton Corp.

Regency Electronics Inc.

Samhill Enterprises, Inc.

Acknowledgments

Security Research International

Shakespeare Co.

Sony Consumer Products

Southern Bell Telephone Co.

SPS Industries, Inc.

Taylor Lock Co.

Technicom International, Inc.

Telco Products Corp.

Uniden Corp. of America

Unisonic Products Corp.

U.S. Tron

Valor Enterprises, Inc.

Webcor Electronics, Inc.

Zoom Telephonics

Contents

CHAPTER 1

CHAPTER 2

Contents

CHAPTER 3

CHAPTER 4

the Bathroom — Add-On Units — Rotary to Pushbutton Dialing —
Add-On Dialers — Advantage of Memory — Telephone Ringer —
Hold Control — Telephone Amplifier — Soft-Tone Add On — Phone
Call Recorder — Time and Temperature Telephone Advertising System

Contents

CHAPTER 9

CHAPTER 10

TELEPHONE BASICS

A telephone is a simple device and its main function is to convert sound energy to its electrical equivalent, or to change that equivalent back to sound again. In doing so it belongs to a family of devices known as transducers.

TRANSDUCERS

A transducer is any component that changes one form of energy to (one or more) other forms of energy. An electric light bulb is a transducer, converting the electrical energy it receives into light energy, but also into some heat energy as well. A toaster is a transducer, changing electrical energy to heat energy, but producing some light energy in the process. A battery is a transducer, and so are photocells, motors, generators, phono cartridges, speakers, and microphones.

Of this group, microphones and speakers are of interest for it is these transducers that help make telephones possible. Microphones are used to convert sound to voltage; speakers (and earphones) for converting voltage to sound.

The Simplest Telephone

The simplest telephone consists of a pair of tin cans, each of which has had its top (or bottom) removed. The cans are connected by a dry string pushed through a hole in the center of the bottom of the can, and knotted to keep it from coming out. The cans must be held, one at each end of the string so the string is taut. At a distance of about 20 feet, someone speaking into one of the cans can be heard by a person holding the other can to an ear, as indicated in Fig. 1-1.

This telephone system, and that is what it is, depends solely on voice energy, and requires no other external energy source such as a battery. The bottom plate of the tin can, vibrating because of sound pressure, transmits these vibrations via the string, to the bottom of the other can, causing it to vibrate. This produces changes in air pressure and it is these changes that results in the sensation we call sound.

This type of telephone has three components: a transmitter which also functions as a receiver; a receiver which also works as a transmitter; and a connecting element, in this case a length of string. So this is not just a telephone but is a telephone system.

This system has its advantages. Its cost is close to zero, it is easily replaced, is practically foolproof, has no patent restrictions, and should have a long working life. But its disadvantages far outweigh these advantages. It has a limited working distance, can only be used in simplex (that is, only one person can talk at a time), produces weak, poor-quality sound, has no way of signaling from one end to the other, and is limited to single person-to-single person communi-

Fig. 1-1. The simplest telephone—two tin cans and a connecting string.

cations. Both parties must be ready at the same time, by prearrangement. And there is no way of modifying it into a practical telephone system.

The Microphone

Without the microphone the telephone could not exist. It is the microphone, working as a transducer, changing sound to an equivalent electrical energy, that makes our telephone system possible. The reason for this is that we can control this energy, make it travel great distances at incredible speeds, route it, examine it with instruments, measure it, and easily convert it to another form of energy, such as sound.

The Carbon Microphone

The amount of current flowing through a conductor, such as a metal, depends on two factors. One of these is the amount of voltage we apply. The higher the voltage, the greater the amount of current that will flow. The other is the resistance of the wire, or other device. The greater this resistance, the smaller the current, assuming we keep the voltage constant. Resistance is measured in a unit called the *ohm*, and voltage is measured in a unit known as the *volt*. With these two we can control the amount of current flow. With voltage we have the added attraction that it can also control current direction.

Thus, we have two ways of controlling an electrical current but in the telephone system it is resistance we use as the controlling element.

Fig. 1-2 shows the basic arrangement of a simple carbon microphone. It consists of a current source such as a battery, some connecting wires, a small cylinder containing granules of carbon, and a flexible metallic plate or diaphragm. In this arrangement there is a flow of current from the battery to and through the carbon granules which have a certain amount of resistance or opposition to the flow of current.

When you use a telephone the only part of the microphone you see is the cover plate of the mouthpiece. This plate has a number of holes in it to permit the passage of sound energy into the microphone. The transmitter (microphone) is positioned directly behind the plate and consists of two parts: a diaphragm and a carbon chamber. The diaphragm is made of extremely thin aluminum with its outer edge firmly attached to a circular frame, somewhat similar to the way in which a drum head is mounted on a drum. While the circumference

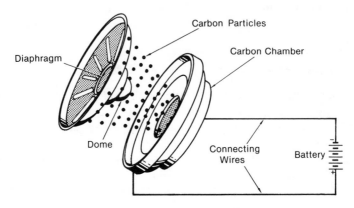

Fig. 1-2. Mouthpiece (transmitter) of a telephone. Although a battery is shown here, it simply represents the dc voltage supplied by the telephone company.

of the diaphragm is fixed, the rest of it is free to move and it does so in a back and forth manner.

On the underside of the diaphragm is a small gold-plated brass dome which nests or fits into a chamber containing carbon granules.

During the time the telephone is in its cradle the transmitter is inactive since the battery is disconnected from the carbon chamber. When the handset is lifted, though, the carbon chamber receives a flow of current.

The highly flexible lightweight diaphragm, held in position by a ring of metal around its circumference, is in direct contact with the carbon granules. When you talk facing the diaphragm, the sound pressure of your voice pushes the diaphragm inward, and in so doing, compresses the carbon granules, reducing their overall rsistance. The louder you talk the greater this compression and the lower the resistance. When you stop talking the diaphragm resumes its usual position, barely touching the granules.

What we have here is an arrangement in which the resistance of the cylinder containing the granules is made to vary in step with changes in voice strength, but in doing so, alters the strength of the electrical current. In effect, we have used sound energy, supplied by the voice, to control electrical energy.

The next step is to take the electrical energy out of the device and put it to work. We can do this as shown in Fig. 1-3. Here we have a *transformer,* a device that has a pair of windings, primary and secondary. A current flowing through the primary winding produces a

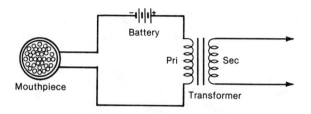

Fig. 1-3. The action of the mouthpiece results in a varying current. In flowing through the primary winding it produces a varying voltage across the secondary of the transformer.

voltage across it. This voltage can be transferred to the secondary winding whenever the current through the primary is made to vary. Whenever that current remains unchanging there is no voltage transfer.

The current through the primary keeps changing whenever sound energy moves the diaphragm. And when this happens we get a varying current in the primary, and a varying voltage across the secondary. But this varying voltage is an analog, a replica in electrical form, of the sound of the voice.

We can now take this varying voltage and use it to produce a correspondingly varying current. And because we use this arrangement to send this current through a pair of wires, we call it a transmitter. The transmitter in this case is the carbon microphone.

The Dynamic Transducer (The Receiver)

All telephones are two-way telecommunication devices, for they can not only transmit, but receive as well. In a telephone a receiver could consist of a coil of wire, mounted near a flexible metal diaphragm. When an electrical current, corresponding to sound, flows through the coil that coil becomes an electromagnet, attracting the diaphragm and causing it to vibrate in step with the varying current. As the diaphragm moves back and forth it produces corresponding changes in air pressure which our ear-brain mechanism interprets as sound. Fig. 1-4A shows the arrangement of the dynamic transducer. This transducer is the receiver in the telephone.

Fig. 1-4B shows the simplest possible telephone system and, except for the conversion of the voice to a varying electrical current, it has a close resemblance to the toy phone of Fig. 1-1.

When the telephone is not in use a steady direct current, supplied

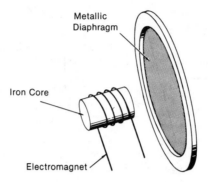

Metallic Diaphragm

Iron Core

Electromagnet

(A) Electrical currents corresponding to variations in the voice are made to flow through a coil of wire, making it into an electromagnet. This attracts the diaphragm to a greater or lesser degree, depending on the strength of the current. The coil is wound around a core of soft iron.

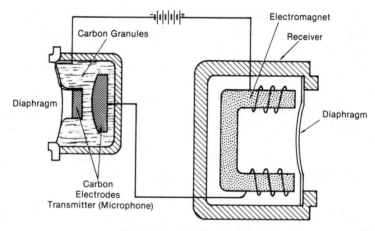

(B) Basic telephone system.

Fig. 1-4. A basic telephone.

by the battery, flows through the transmitter (microphone) and through the receiver. When a person speaks into the microphone the flexible diaphragm applies a varying pressure to the carbon granules, causing its resistance to change in step with the sound. This varying current flows through the coil windings of the receiver, making it into an electromagnet. Since the current is being made to vary by the voice the electromagnet's strength also varies in step, attracting the flexible diaphragm to a greater or lesser amount. The moving diaphragm changes the air pressure around it, producing sound.

As shown, the setup in Fig. 1-4B is a one-way arrangement only. It can be made two way by duplicating the equipment using a receiver and transmitter at both ends.

Another requirement, before this can be made into a practical telephone system would be some kind of switching arrangement so as to connect telephones to each other.

The Handset

Telephones can be one-piece or two-pieces device. In a two-piece telephone, as in a candlestick type, the receiver and the transmitter are housed separately. Fig. 1-5 shows a two-piece telephone.

One-piece telephones are much more commonly used. These, as illustrated in Fig. 1-6, house the transmitter and the receiver in a single frame. Not only is the single-piece telephone more compact, but there is less possibility of audio feedback, a condition that results in a howling sound. Audio feedback is also known as acoustic feedback.

THE CENTRAL OFFICE

To be able to communicate, a telephone must be connected to other telephones, whether they are nearby or at some distance.

Fig. 1-5. The two-piece telephone.
(Courtesy Comdial Corp.)

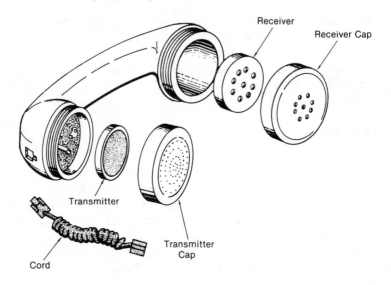

Fig. 1-6. The one-piece telephone handset.

Basically, then, all telephones must be connected to each other, not simultaneously, but selectively.

In Fig. 1-7 there are five telephones. These can be connected directly by pairs of wires, but to do so requires 10 pairs of lines so that any one of these telephones can be connected to any other phone. However, by using some type of switching device, either mechanical, electromechanical, or electronic, these same five phones can interconnect using only five pairs of telephone wires, called lines or loops. The switching is done automatically by a switching device housed in a Central Office operated by your telephone utility.

Switching

In telephone language the act of switching means making a connection. When a pair of telephones, one making the call, the other receiving it, are switched, they are connected.

There are two basic types of switching — manual and automatic. The Central Office uses automatic switching while manual switching is the kind often used in business offices. If the office has a telephone operator then the call can be manually switched to any one of a number of telephones.

Not all utility Central Offices use the same kind of switching equipment. There are three types: step-by-step switching, crossbar

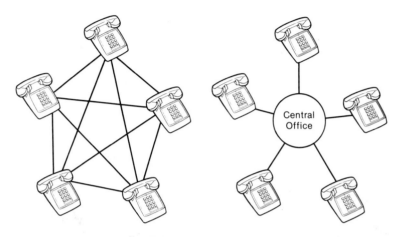

Fig. 1-7. Without Central Office switcher directly connected telephones require 10 pairs of interconnecting lines (left). With Central Office, number of connecting lines is reduced by 50%.

switching, and electronic switching. Step-by-step and crossbar switching are gradually being phased out in favor of the quicker, non-mechanical electronic switching.

Fig. 1-8 shows a basic arrangement of a telephone call between a pair of subscribers both of whom use the same Central Office of the telephone company. Subscriber A picks up the telephone and is notified by a dial tone that a clear line is available to the switching system of the utility. At the utility, the switching system automatically connects the call to subscriber B, notifying the subscriber by a ringing current which operates a bell inside the telephone. When subscriber B picks up the telephone the connection is completed.

When subscribers are serviced by different Central Offices, the connection between the pair of telephones is via a trunk cable, as indicated in Fig. 1-9. The call is received, in this example, by subscriber A's Central Office. The call is then automatically routed to a Tandem Office and from there to subscriber C's Central Office.

TELEPHONE LINE VOLTAGE

The source voltage for your telephone is 50 volts dc and is supplied by your local telephone company. However, when you use your telephone, this voltage drops to about 4.6 volts. That is for one tele-

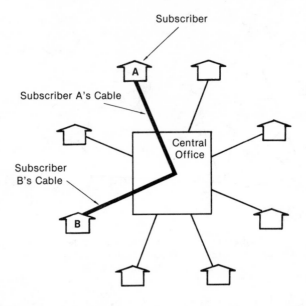

Fig. 1-8. When a pair of subscribers use the same Central Office, there is a direct wired connection between subscriber A to the Central Office and also a similar connection between subscriber B and the same Central Office.

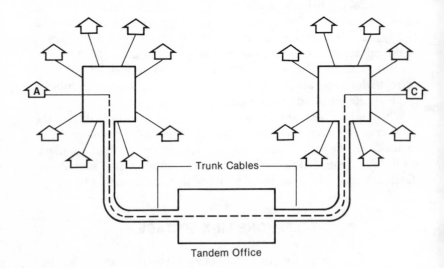

Fig. 1-9. When a pair of subscribers do not use the same Central Office, the call is routed via a supplementary switching center called a Tandem Office.

phone. If you use two telephones the voltage decreases to 3.2 volts, while a third telephone drops the voltage to approximately 2.3 volts. In short, you cannot keep adding telephones indefinitely. To know how many telephones you may have in your home, or in your business, check with your local telephone utility.

THE DIAL TONE

When you put the receiver of a telephone to your ear, the action of removing the telephone from its supporting cradle automatically closes a switch that lets you hear a dial tone. The dial tone is an audio current, a signal from the Central Office that the two signal-carrying wires from your phone are now connected to their switching system.

Your telephone does not contain a battery nor is it connected to the ac power line. But it must have an electrical current so as to function and it gets this current from the phone utility's Central Office. The amount of this current varies with your distance from that office. The greater the distance, the higher the resistance of the line, hence the smaller the current. This current can vary from as little as 20 milliamperes (20 thousandths of an ampere) to as much as 80 milliamperes. As indicated earlier, the voltage used by the Central Office for producing the current is 50 volts dc, an amount that together with the maximum available current is too small to present a shock hazard.

HOW YOUR TELEPHONE CALL TRAVELS

The call you make on your telephone is made via connecting wires (also called a loop or line) and can include radio transmission for long distance. When you initiate a telephone call, your voice, changed to a corresponding electrical current by a microphone in your telephone, travels through wires to a Central Office. Switching equipment in that office sends the current on its way to the called telephone. At the receiving telephone the electrical current is converted by the receiver (the dynamic transducer) to sound.

As the electrical audio current travels through wires from your telephone to a Central Office for switching that current gradually becomes weaker. For long distances repeater stations are used. These are stations equipped with solid-state audio amplifiers which strengthen the audio current.

LONG DISTANCE TRANSMISSION

The technique of connecting individual telephones to a Central Office by using a pair of wires for each telephone message is impractical for long distance communications. Instead, it is possible to send 10,000 messages or more along the same telephone line by using a process known as carrier transmission and also by using special wires known as coaxial cables.

An alternative way is to employ microwaves as a form of radio transmission using ultrahigh-frequency waves having the characteristic of traveling through space in straight lines. Such waves can also be concentrated into a narrow beam. In this way they can be radiated from one relay station to another, with the relay stations positioned about 30 miles apart.

Each relay station has a receiver for picking up the ultrahigh-frequency (uhf) waves from a preceding relay station, and a transmitter for sending them along to the next station. A microwave system has the capability of handling about 36,000 telephone calls.

Another technique known as over-the-horizon radio relay system aims the microwaves at the horizon instead of the next relay station. In this way relay stations can be separated by distances of up to 200 miles. The advantage is that it eliminates the need for so many repeater stations. However, in the over-the-horizon technique the signal is much weaker when picked up by the repeater, so the receivers at such stations must be more sensitive, while the transmitter portion must be more powerful.

PARTS OF A TELEPHONE

The essential parts of a telephone are the transmitter and the receiver, but by themselves, they are not enough. You must have some way of knowing when someone is trying to talk to you and so the telephone must be able to alert you that you should use it.

The telephone also needs a switch to be able to connect and disconnect it from the loop. It must have some kind of device to be able to separate the incoming electrical signals from those that are outgoing. And it must have some way of notifying the Central Office as to the number you have dialed.

The Varistor

The strength of the sound you hear from the receiver in your telephone depends on the amount of loop current supplied by the utility's Central Office. If, as mentioned earlier, you are close to that office, the loop current will be larger than if you were located farther away. As a result the sound produced by your telephone receiver will be very loud. Conversely, at some distance from the Central Office the sound could be weak. To compensate for this deficiency telephones are equalized with the help of varistors. A varistor is a kind of variable resistor. It has a low resistance for long loops and a much higher resistance for short loops. Its effect is to produce an average loop current that is much the same for short or long loop lines.

Sidetones

Whenever you talk to someone directly, instead of using a telephone, the first person to hear what you say is yourself, even if you are standing close to some person listening to you. The distance between your mouth and ears is quite short and so you hear what you say. This is essential since it enables you to control your voice level. People who have hearing problems tend to talk more loudly than those who have good hearing.

But when you talk on the telephone one of your ears is covered and so, in effect, you cut your hearing ability by 50 percent, assuming you hear equally well with both ears. To overcome this, some of the voltage produced by the transmitter is routed to the receiver. When you use a telephone you also talk to yourself.

The sound current that is sent to the receiver from the transmitter in the telephone is referred to as a sidetone. The amount of sidetone current is very carefully adjusted so as to enable you to speak at the same voice level you would use without the telephone.

The Hookswitch

Every telephone, except handsets used with wireless telephones, is equipped with an automatic switch. When you lift a telephone from its cradle you automatically turn this switch to its on or closed position. This action connects your telephone to the loop and current flows from the Central Station to your telephone. The current flow closes a circuit in the Central Office that, in turn, connects your loop to a circuit that supplies a dial tone. This dial tone is your indication that your connection to the Central Office has been completed, is functioning, and that you can proceed with dialing. In the event there is no dial tone, dialing will be useless.

What Happens When You Dial?

As an example, assume you plan to dial a number such as 722-7562. The first three digits identify the particular Central Office to which you want your telephone connected. Each Central Office has a different location.

If you dial a number having seven digits only, then the absence of additional digits indicates the call is a local one. For long-distance calls ten digits are needed. The first three digits indicate the area. Thus, if the first three digits are 201 this indicates the Central Office is located in New Jersey; 212 signifies the Central Office is in New York. Thus, whether you use the first three numbers for a local call, or the first six digits for a long distance call, these are simply for narrowing down the area of location of the Central Office. It is only the last four numbers that identify the particular telephone to which you want to be connected.

Rotary and Pushbutton Dialing

The telephone you use may be equipped with a rotary dial or may be a pushbutton unit. Assume you have a rotary dial and that the first number you dial is the digit 5. In dialing, your first step is to turn the dial to digit 5 and then release the dial. When you do this it operates a mechanism which interrupts the circuit five times. Each interruption produces pulses at the input equipment of the Central Office. And so when you dial 5 you produce 5 pulses. When you dial 6 you produce 6 pulses. If you use pushbutton dialing pulses are also produced with the Central Office equipment recognizing each tone as a number of pulses.

The first group of three pulses (for local calls) or the first group of six pulses (for long-distance calls) connects your telephone either directly or indirectly with the correct Central Office. The last group of pulses connect your telephone, on a step-by-step basis, to the particular telephone you want to reach. When this happens, the circuit to that telephone is closed, and a ringing current is automatically sent over that line. That ringing current, operating a bell or chime in the called telephone, will continue until the handset of the called telephone is picked up.

The Phone Ringer

Just about all telephones have a ringing device as an alert that a call is coming through. There are some plug-in type extension phones that do not have a ringer, but instead have a flashing light to call attention to the call. There are also some telephones that are equipped with a bell plus a flashing light. The user has the option of having either the bell or the light, or both, operative.

The usual telephone has a gong or bell and sometimes a pair of these. The ringer circuit (Fig. 1-10) consists of an armature, an iron-core coil, a source of ringer current, a spring, a bell, and a clapper. The armature is a length of ferrous metal and is pivoted at one end. When current flows through the iron-core coil it becomes an electromagnet and attracts the armature. This action opens the circuit and current flow through the iron-core coil stops. Since it is no longer an electromagnet the coil cannot hold the armature which is pulled back into its original position by the spring. When this happens the clapper strikes the bell.

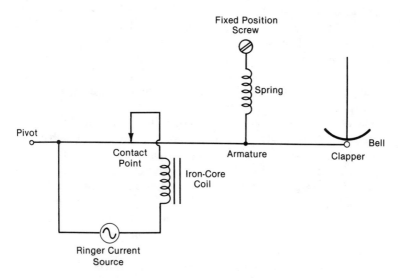

Fig. 1-10. Basic concept of a bell ringer. The ringer current source is low-frequency ac, supplied by the telephone utility.

At this time the contact point touches the armature and current flows once again through the iron-core coil. This attracts the armature and pulls the clapper away from the bell. However, the circuit is opened at the contact point once again, and the spring pulls the clapper against the bell.

The ringing action will continue until someone at the called telephone lifts the receiver. This action operates a switch which stops the flow of ringer current.

PULSE FORMATION

The current from a battery is dc, an abbreviation for direct current, an indication that this type of current always flows in the same direction, from the minus (negative) terminal of the battery, through some external device (often called the load) and then back to the battery + terminal again. Having arrived at the plus terminal (positive) the current continues its flow through the internal structure of the battery, and reaching the minus terminal continues on its path again. Actually, this description of current flow is highly simplified, for we cannot say that current starts at any particular point, any more than we can say that any point on a wheel is the beginning of rotation of that wheel.

Fig. 1-11 shows the graph of an electrical current flow plotted against time. When the switch of a dc circuit is closed, current rises to its maximum value and this rise time can be so rapid that for practical purposes we can call it instantaneous. If no changes are made in the circuit or load, current flow continues until a switch is opened and the current drops to zero.

The graph shown in Fig. 1-11 can be called a pulse and in this case represents just a single pulse of current. But we need not have just one large pulse but can have an entire series of pulses as shown in Fig. 1-12. The number of pulses we have per second (pps) is established by the rapidity with which we operate the on/off switch. Thus, we could have 10 pps, 15, or 20, or any number. If we operate the on/off switch with some precision we can make sure that each pulse has the same time duration, hence the same width, as any other pulse. The time duration of a pulse is known as its on time, and the time duration between pulses is referred to as off time. In pulse formation it is desirable to have all on time pulses the same, and also to have all off time segments the same. The ratio of on time to off time

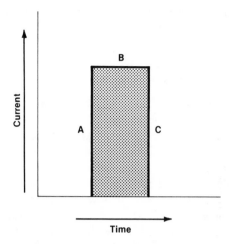

Fig. 1-11. Graph of a pulse of current. The current rises to its maximum value almost immediately, represented by the vertical line at the left (A). The current continues flowing (B) but when the switch is opened, the current drops to zero (C).

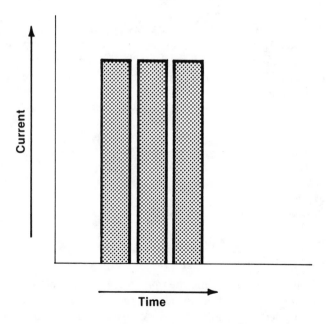

Fig. 1-12. A series of pulses can be produced by turning a switch on and off.

can be 1 to 1, 2 to 1, 3 to 1, or any other ratio we want. While this may be difficult to do manually, it is much easier to have fairly precise on and off times electronically.

Current Pulses vs Voltage Pulses

The pulses shown in Fig. 1-12 are pulses of electrical current. We can convert these current pulses into equivalent voltage pulses in a number of ways. One of the easiest is to send the current through a component known as a resistor. As the current flows through the resistor, a comparable voltage develops across it. If we have several resistors, connected in sequence, a voltage will be produced across each of the resistors. If the current flowing through these resistors is in the form of pulses, that is, current that is turned on, off, on, off, etc., then the voltage produced across the resistors will have the same shape, that is, will be pulses of voltage.

THE ROTARY DIAL

While there are several rotary dial styles and sizes they are basically all alike. All of the numbers except 1 and 0 are accompanied by letters. These numbers and letters can be placed inside the dialing holes or outside it and immediately adjacent. The letter and number arrangements are:

```
1
2  A B C
3  D E F
4  G H I
5  J K L
6  M N O
7  P R S   (Letter Q is not used)
8  T U V
9  W X Y   (letter Z is not used)
0  Operator
```

Digit 0 is also used for dialing the operator. Dialing a number or any of the accompanying letters produces the same number of pulses. Thus, dialing letters D, E, and F is exactly the same as dialing 3.

Fig. 1-13 illustrates typical rotary dials. The dial is rotated

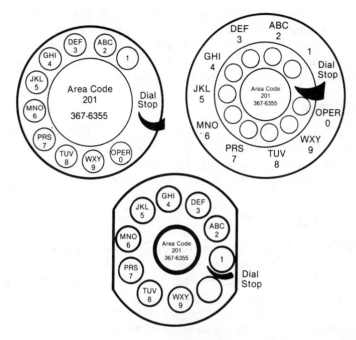

Fig. 1-13. Various styles of rotary dials. The number and lettering system is the same for all. The metal stop is the limit of travel of the dial when it is rotated clockwise.

clockwise, but since it is spring loaded always returns to its starting position. The numbers are arranged so digit 0 is closest to the starting point, followed by digits 9, 8, etc. Note also, that in rotating the dial, digit 0 requires the longest time to reach the maximum dialing position. This maximum dialing position has a limit established by a metal stop across the dial.

While the dial is spring loaded, using just a simple spring would mean that the dial would rotate more rapidly on its return after the 0 position on the dial had been used. Thus, if you insert your finger in the 0 hole on the dial and rotate the dial until you reach the stop, the spring would have its maximum elongation and the dial would return extremely rapidly. On the other hand, if you dialed digit 1, the travel of the dial would be small, the spring would be barely pulled, and the dial would return comparatively slowly.

To avoid this kind of spring action the dial is equipped with a governor. The governor controls the rate of return of the dial, making its motion such that it is the same no matter which number is dialed. It does not help to put your finger in one of the dial holes and try to force the return to act more rapidly, since the return action will be controlled by the governor anyway. Forcing the dial return can sometimes damage the spring and the governor.

Dial Speed

The usual speed of return rotation is 10 pps. This is the rate no matter which number is dialed. Thus, dialing the number 5 five pulses of current, but at a rate of 10 pulses per second. If you dial the digit 0 you produce ten pulses of current, but once again, at a rate of 10 per second.

Interdigit Time

When you use a rotary dial you produce a series of pulses corresponding to the number on the dial. But you do not move from one digit to the next instantaneously. The time between each group of pulses is referred to as interdigit time.

When you write a series of numbers on a sheet of paper you allow some space between the digits, or you may use a comma to separate the digits into groups of three. This makes it easier for you to read the numbers. Similarly, interdigit time is needed by the switching equipment at the Central Office. Each series of pulses represents some number in the decimal system, numbers ranging from 0 through 9. Since all the pulses are of the same strength, and since they all have the same spacing, interdigit time is essential to keep one group of pulses from running into another.

Interdigit time is very short, and is less than one second, ranging from a half second (500 milliseconds or 500 msec) to about 700 milliseconds.

Switching Time

A switch that is closed is said to be in the *make* position. Conversely, an open switch is in the *break* position. The time that a pulse is on, or in its *make* mode is 39 percent of the entire time of a single pulse. Consequently the off or *break* time is 61 percent. The sum of these two numbers, 39% + 61% = 100%, is the total duration of a pulse, including its make and break time. The percent break ratio deliberately favors off time for this represents the most economical use of battery power supplied by the telephone utility.

Touch-Tone® System

Touch-Tone® is a trademark registered by the American Telephone and Telegraph Co. (AT&T) and this is the copyrighted phrase now used by the Bell System. It is a signaling system now being more and more extensively used and which may ultimately replace the rotary dial method. Independent manufacturers of telephones using the Touch-Tone® method often designate their dialing system by some coined name such as Pushbutton Dialing, Tone Fone, Quick Touch, or Soft Touch. They all work in the same way by sending a series of tones to the Central Office of the telephone utility instead of a series of pulses.

This type of dialing is also known as *dual-tone multifrequency dialing* (DTMF) since depressing any single button produces not one, but a pair of tones of different frequencies.

The advantage of DTMF is that it permits faster dialing. There is no waiting time, as in the rotary dial method, for a dial to resume its starting position and you can dial a number just as fast as you can move your fingers. However, even with DTMF we have the equivalent of interdigit time, for it does take some time, however short, to move from one pushbutton to another and to depress it. The tones that are produced have a frequency that is precisely determined. The pairs of tones for each digit must not only be accurate but must be present simultaneously to be used by the switching equipment in the Central Office.

Tones

The two tones used in pushbutton dialing are in the audio-frequency range and are categorized as low and high. There are four tones in the low group and three in the high group. The low tones are 697, 770, 852, and 941 hertz (cycles per second, or cps), while the high tones are 1209, 1336, and 1477 hertz. Note these tones are not harmonically related. For example, the second harmonic of 697 Hz is $697 \times 2 = 1394$ Hz. While this harmonic falls within the frequency range of the high tones, it is situated somewhere between the top two of the high tones, 1336 Hz and 1477 Hz. The harmonics of the remaining low tones fall outside the high tone range. Thus, the second harmonic of 770 Hz is 1540 Hz. Consequently, there is no harmonic interference between the tones.

Tone Digit Assignment

Instead of having pulses assigned to each digit, the tones are supplied according to the following table.

Table 1-1. Tone Digit Assignment

Dial Digit or Symbol	Low-Tone Frequency (Hz)	High-Tone Frequency (Hz)
0	941	1336
1	697	1209
2	697	1336
3	697	1477
4	770	1209
5	770	1336
6	770	1477
7	852	1209
8	852	1336
9	852	1477
*	941	1209
#	941	1477

There are two other designations on the pushbutton pad, other than the digits. The symbols used are the asterisk (*) and the space mark (#). The tones used for the asterisk are 941 Hz and 1209 Hz and for the space mark are 941 Hz and 1477 Hz. The tones used for the digits, asterisk and space mark have frequencies that are not arbitrarily selected by the Bell System or any other operating telephone companies but are set internationally. The frequencies that are used must be accurate to within 1.5.%

Except for physical design the pushbutton telephone pads have the same arrangement as rotary dial types. But there are a few differences. A rotary dial has a total of 10 operating positions; a pushbutton type has a total of 12. The two extra positions are those represented by pushbuttons marked * and #. Sometimes pushbutton #1 simply has the number 1 and no letters while with other telephone pads the digit 1 is accompanied by the letters Q and Z. These are the letters that are omitted from the rotary dial.

PULSE-TYPE PUSHBUTTON SYSTEM

So far we have two distinct types of signals. One of these is rotary dialing, a system that produces a series of pulses. The other is the

Touch-Tone® method, a system that produces a series of pairs of tones. It is also possible to have a combination of the two, often referred to as a pulse-type pushbutton system. This method makes use of pulses, just as a rotary dial type makes use of pulses, but instead of a rotary dial, a pushbutton pad is used instead.

Not all Central Offices of the various telephone companies are equipped to work with dual tone multifrequency signals and can accept pulse signals only. There are phones available which have a pad using pushbuttons, but these produce pulses instead of tones. This type is not as fast as the true dial frequency phone type, but are faster than rotary dialing. As phone companies acquire equipment to handle tone dialing, these phones and the rotary type will possibly be phased out.

The tones produced in the telephone are generated by a solid-state oscillator using power supplied by the telephone company Central Office. An oscillator is an electronic generator producing a small ac voltage. Fig. 1-14 shows how to determine the frequency of any pair

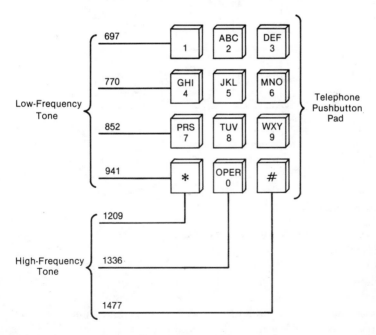

Fig. 1-14. Pairs of tones corresponding to pushbuttons on a telephone pad. Each button produces a pair of tones. The numbers represent the frequency in Hertz (cycles per second).

of tones corresponding to any digit on the telephone pad (Fig. 1-15). To learn what these tones are, put your finger on any number and move directly downward to locate the high tone. Put your finger back on the number and move to the left to determine the low-frequency tones.

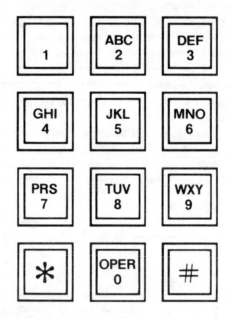

Fig. 1-15. Pad for a pushbutton-type telephone.
(Courtesy Southern Bell Telephone Co.)

Touch Sensitive Dialing

While the most common dialing methods are the rotary dial and pushbutton, there are other methods. One of these makes use of a solid touch-sensitive surface imprinted with the keyboard positions. These touch buttons are overlaid with a printed sheet of paper which outlines the buttons. The printed sheet of paper can indicate the usual dialing numbers and special keys utilizing symbols such as # and *. Some telephones also use touch-sensitive surfaces for features such as names in storage. An advantage of the touch-sensitive type is that it is much faster than using a rotary dial and quite possibly the pushbutton type. However, it is susceptible to more dialing errors.

TELEPHONE MNEMONICS

A mnemonic is any memory aid. If you have trouble remembering a particular telephone number or if you want people to be able to keep your telephone number in mind, you can use a mnemonic to make remembering easier. A way of doing this is to make use of the letters that accompany the numbers on a telephone pad or rotary dial.

If you will look at the dial or pad you will see it consists of digits and letters. Digit 2 is also ABC, 3 is DEF, and so on. If your telephone number is 884-2628 this could be remembered as *TUG BOAT*. If your number is 937-6226 it could be remembered as *YES MAAM*. Unfortunately, not all phone numbers lend themselves to mnemonic use. However, each number does have three letter possibilities.

THE TELEPHONE EXCHANGE

A telephone can send pulses of current, or a series of paired tones, to a Central Office. When this happens, the Central Office must route these to the correct called number. The device or mechanism that connects your signal with a called telephone is known as a *telephone exchange*.

TELEPHONE POWER

Your local telephone utility does not depend exclusively on the ac power line for its source of power since doing so would mean failure of the entire telephone system in the event of a power outage. Instead, the electrical energy utilized by the Central Office consists of lead-acid batteries and while these operate on the same principle as a car storage battery they are much more substantial. They consist of four 12-volt units wired in series to supply a total of 48 volts. The voltage, of course, is dc, as are all battery voltages. This battery system is shunted by a trickle charger, a device that rectifies and filters current supplied by the ac power line. In this way the Central Office's battery system is always kept at full charge so even if the power line does go out telephone service will not be affected.

The battery voltage used by your local telephone utility is variously referred to as 48 volts or 50 volts. Each of the fully charged batteries is 12.6 volts and since there are four of these in series the total is 4 ×

12.6 or 50.4 volts. This voltage remains fairly constant, but it can decrease slightly depending on the load.

This is the amount of voltage that appears at your telephone when that phone is not being used. However, the moment the telephone is activated, there is a voltage drop across the telephone lines carrying current to your phone. And, as indicated earlier, this is just a few volts.

THE RINGING SIGNAL

The ringing signal you hear as an alert that your telephone is on a called circuit is initiated at the Central Office. It can be produced by a device known as a ringing generator, but, in more up-to-date arrangements, by solid-state circuitry. The advantage of solid state is that unlike the ringing generator it has no moving parts and has a longer operating life with less downtime.

As shown in Fig. 1-16 the wires connected to the ringer circuit consist of a pair coded yellow and red. The iron-core coils shown are part of the ringer. When these coils are energized by a ringing current, they operate an armature, which, in turn, actuates the clapper of the ringer as indicated earlier in Fig. 1-10. Connected between the ringer coils (L1 and L2) is a capacitor (C1). The coils (inductors) plus the capacitor, form a tuned circuit, much like the tuned circuits we have in radio and television receivers.

IMPEDANCE

All devices connected to a telephone have a certain amount of impedance. The higher the impedance of a telephone or a telephone accessory the smaller the amount of current it demands from the telephone line.

There are two impedance extremes. One is a short circuit and represents minimum or zero impedance. Because it is zero impedance or very close to it, there is a heavy demand for current, representing a heavy load. The opposite is an open circuit, possibly represented by an open switch. This is maximum impedance. No current flows and the load is considered to be nonexistent. Most impedance conditions are between these two extremes. Sometimes terms such as low impedance and high impedance are used but these are only relative.

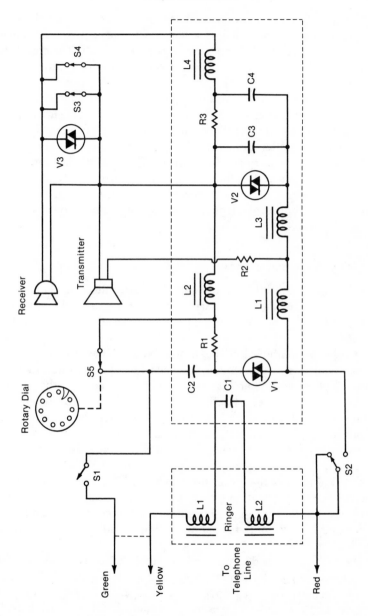

Fig. 1-16. Rotary dial telephone circuit. The green and red wires connect to the telephone lines for receiving and sending voice signals. The ringer current is supplied by the yellow wire.

All telephones and all accessory telephone devices connected to the telephone lines are regarded as a load on the line. Any component that requires a relatively large current is a heavy load. The corollary is that a component that demands a small amount of current is a light load.

As long as a telephone is not being used it acts like an open circuit, that is, a device having an infinite amount of impedance. However, if there are two or more telephones in your home or business, they are all wired in parallel, sometimes called a shunt connection. This means if one of them rings, they all ring. The total ringing current is equal to the sum of the individual ringing currents for each telephone. This means the total impedance is lower. Consequently, you should notify your local telephone company if you plan to use two or more telephones.

Ringer Circuit Impedance

The ringing coils (Fig. 1-16) and their associated capacitor (C1) comprise what is known as a series-resonant circuit, a circuit that is characterized by having minimum impedance at its resonant frequency.

The series combination of L1, C1, and L2 has an impedance, which, while low at the ringing frequency, can be measured by your local telephone utility, and it can do so at any time since the ringing circuit must be connected across the telephone lines (coded red and yellow in Fig. 1-16). Some telephone utilities monitor this line constantly and are immediately made aware if there is any change in impedance. This can happen, for example, if you add more telephones to your existing telephone lines. The telephone company then has the right to ask that you supply them with the ringer equivalence and the FCC authorization type number. Further, they have the right to charge you for the use of this additional equipment. However, you can connect *receive only* telephones without making any connections to the ringer circuit.

The current supplied to the ringer circuit is not dc but is ac, not the ac supplied by your local power company which is generally 50 Hz or 60 Hz, but a lower frequency. This lower-frequency ac voltage is produced by the lead-acid batteries, mentioned earlier, starting as dc and then converted to ac (actually pulsating dc). Capacitor C1 in Fig. 1-16 performs a double function. It helps tune ringing coils L1 and L2 and it also works to block any dc that might accidentally be routed into the ringer lines. A capacitor will pass a varying current, such as the ringing current, but will block the passage of any direct current.

SWITCHING ARRANGEMENT

Fig. 1-16 shows a number of switches identified as S1, S2, S3, etc. As shown in this diagram, S1 and S2 are open, while the other switches (S3, S4 and S5) are closed. This is the state of affairs when the telephone is not being used and is resting in its cradle. Note that the ringer circuit is not affected by any of the switches and is across the telephone lines at all times.

With the telephone in its cradle, switches S3 and S4 are closed, thus acting as a short across the receiver and the transmitter. When the telephone receives a ringing signal and the telephone is lifted from its cradle (S3 and S4 open) the short across the transmitter and receiver is removed. At the same time switch S1 becomes closed, thus connecting the rotary dial directly to the telephone line, coded green. Switch S5 also opens at the same time, permitting pulses to be controlled by the rotary dial. Note that all of these switches, S1 through S5 inclusive, are part of the hookswitch. This is the only switch you can actually see on a telephone and it is the switch you sometimes jiggle when you do not receive a dial tone or you want to attract the attention of the operator.

When the telephone rings and you pick it up, switch S3 remains closed until S1 and S2 have closed connecting the telephone to the line. Thus, there is a time delay between these switches, but once S3 is opened the receiver of the telephone is ready for action.

MAKING A CALL

Now, assume that instead of using the receiver, you want to initiate a call. Lifting the telephone from its cradle automatically produces a dial tone, supplied by the Central Office, indicating there is a clear line to the exchange. Before dialing, with the rotary dial in its rest position, switch S5 is closed. The closing of S5 connects the rotary dial to the line. But as the rotary dial is turned, there is a series of circuit openings and closings, taking place at the rate of approximately 10 times per second every time the dial is turned. As indicated earlier, dialing the digit 0 results in 10 switch openings and closings, while dialing 5 produces five such actions.

The opening and closing action of switch S4 is to prevent the clicks caused by the opening and closing actions from being heard in the receiver.

If you will examine S5, you will see that a resistor (R1) is in series with capacitor C2 and that these two components, known as an RC (resistor/capacitor) network is shunted across switch S5. The very rapid make and break action of the rotary dial switching system, as mentioned earlier, is used to produce pulses of current. The result is the generation, not only of the pulses, but of electrical noise as well and this noise can cause interference in nearby radio receivers. The capacitor of the RC network absorbs the electrical noise pulses and the resistor dissipates them. The action of the rotary dial is comparable to that of the sparking produced by the brushes on the commutator of a generator.

INITIATING THE DIAL TONE

Lifting the handset is equivalent to a demand for current. This current demand is a signal to the Central Office switching circuits to supply a dial tone, and this is done automatically.

The ringing circuit of a telephone will continue to function as long as the telephone handset remains in its cradle. When the handset is lifted, the ringing current decreases to the point where it can no longer actuate the gong or gongs and so the ringing stops. When you receive a phone call, and prefer using an extension phone instead, you can hang up without disconnecting the call, provided you do so in 10 seconds or less. However, this feature cannot be used by the party making the call. If they should hang up, deciding to use an extension phone in preference, the call will be disconnected.

ON-OFF LINE SWITCH

For pushbutton telephones the keyboard, sometimes called a phone pad, or simply a pad, consists of 10 numbered pushbuttons, 1 through 0, plus two additional pushbuttons, identified by the symbols * and #. These two pushbuttons, either one or both, can be used for last number redialing. Note that on some keyboards, such as the one illustrated in Fig. 1-17, there is also an on/off line switch. Its function is to prevent accidental cutoff of the telephone, something that

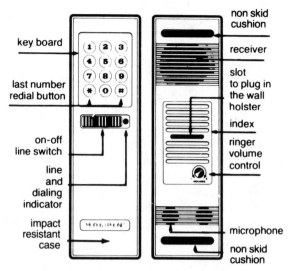

Fig. 1-17. Telephone equipped with on-off line switch.
(Courtesy Modulo-Phone, SPS Industries, Inc.)

sometimes happens when a telephone is handled, possibly in picking up or laying down. This particular telephone also has a line and dialing indicator. As long as the light of this indicator glows, the phone line is active, that is, the telephone line has not been accidentally cut off.

The drawing at the right in Fig. 1-17 shows the rear of the phone. It is equipped with a pair of nonskid cushions. The friction of these cushions helps keep the telephone in position on a slippery surface or if the phone is put down on a surface that is slanted.

EXTENSION TELEPHONES

There are various types of extension telephones. One could be a regular telephone, capable of receiving and initiating calls, complete with a rotary dial or dialing pad. It is sometimes referred to as an extension telephone since it may be in a den or a bedroom. However, it requires a telephone outlet and it incurs a telephone charge. While it is called an extension telephone it is really a supplementary telephone, and sometimes does not have as many features as the main telephone.

Another kind of extension telephone is one that does not have a keyboard. A telephone without a keyboard is an answer only telephone and has a number of advantages. An attractive feature is that it prevents unauthorized phone calls. Thus, a telephone of this type could be used in a children's bedroom. It can be used to answer calls from anywhere in the house, but, like a telephone that is pad equipped, it must be plugged into a telephone outlet. (Fig. 1-18 illustrates an extension telephone.) It can have an on/off line switch to prevent possible accidental cutoff of the phone line. It can also have a line indicator that glows as long as a connection is made.

Still another kind of extension telephone is one that does not have a ringer circuit. It can receive phone calls and you can use it to dial an outside call. An extension phone of this kind is sometimes used when the ringing of the telephone would be too disturbing. However, the main telephone would have a call alert, either a bell or chime.

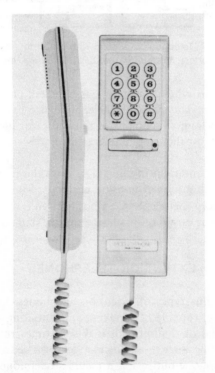

Fig. 1-18. This French telephone is available with or without a keyboard as an extension phone in a child's room or elsewhere.
(Courtesy Modulo-Phone, SPS Industries, Inc.)

ACOUSTIC FEEDBACK

Acoustic feedback, mentioned briefly earlier, is a condition in which sound is fed back from the transmitter to the receiver, and from the receiver back to the transmitter, continuing until there is a howling sound. This is always possible with two-piece telephones when the mouthpiece and the receiver handset are placed too close to each other.

With the single piece telephone, the microphone of the transmitter and the speaker of the receiver are physically separated and cannot be brought together. The distance between them, while small, is enough to minimize the possibility of feedback. Further, the circuit arrangement of telephone wiring is such that the transmit and receive signals cannot cross lines and interfere with each other.

DEDICATED TELEPHONE LINE

A dedicated telephone is one that is used for a single, specific purpose, generally for emergency calling. Thus, a particular telephone and its telephone number could be *dedicated* for calling the police. The telephone is not used for any other purpose. In this way the telephone line is always kept open for this special use.

A dedicated telephone can be used for a bedridden invalid or in a business that requires the utmost in security. It could also be used by children provided they are fully instructed as to why and when the telephone should be used. A dedicated telephone is one that is restricted, and if the phone is programmed, as it should be, then it is not necessary for the full emergency number to be dialed. Instead, the user simply depresses a single pushbutton and the call is made automatically. No conversation is required, since, by prearrangement, the called service can recognize the number doing the calling and will also regard it as a call for help. The called number could be a police department or a hospital emergency service. Alternatively, it could be a relative or a friend. However, the called number should be one that maintains a 24-hour emergency service.

It is important for the emergency telephone to be recognized as such. It should be a distinctive color, possibly red, and it should carry a written or printed sign on it, readily noticeable, as to its use.

DESIGNATION OF AREA CODES

When you make a telephone call, charges do not start until you are connected to the called number with that called number responding. Until the connection is made the telephone company absorbs the cost of operating the phone circuits. Obviously, it is in the phone company's interests to have the call go through as quickly as possible.

For pushbutton telephones, every digit requires the same amount of time. However, with a rotary dial phone, area codes such as 718 require more time than codes such as 212 (New York City) or 213 Los Angeles) or 214 (Dallas) or 215 (Philadelphia). When you use a number such as 8 or 9 on a rotary phone you must wait for the dial to return to its zero or starting position before dialing the next number, something that does not happen with a pushbutton phone. Although the waiting time for an individual phone is small, in the aggregate for all phones it can be quite large.

For that reason telephone companies try to assign low area code numbers to regions of high population density. There are many more telephone calls to cities such as New York, Chicago, and Los Angeles than to a city such as Blue Springs, Missouri or Broken Arrow, Oklahoma. This does not mean that all low population density areas have high area code numbers. Area code numbers were assigned before studies were made of the effect of high area code numbers on calling time. And so, on occasion, a telephone company may try to change an area code to make it correspond more closely to population density and the total number of calls registered.

ABBREVIATIONS

As in many other industries, various abbreviations and acronyms are used in telecommunications.

PAX

Private automatic exchange is referred to as PAX. Not all telephone systems are open to the public and it is possible for a group of subscribers to have their own telephone setup. Those who are members of a private automatic exchange can use their system to communicate with each other, an operation that may be faster and supply more security from line tapping.

PABX

This is an abbreviation for *private automatic branch exchange.* With this system the subscribers can have telephone communications privately, just as in the PAX system. The difference is that with PABX the users can also make telephonic contact with those having a public telephone system.

PBX

Not all offices have a separate switchboard. As an alternative, it is possible to have each individual telephone equipped with a switching arrangement, known as PBX or *Public Branch Exchange.* As an example, a business could have four separate telephone lines, with all of the lines brought into a number of telephones. These telephones are equipped with a series of pushbuttons (in addition to the usual pushbutton dialing pad). Any incoming call can be answered by depressing the correct pushbutton. The telephone is also equipped with pushbuttons for transferring any incoming call to any of the other telephones in the office.

These telephones are also equipped with a double alert. There is a ringing bell as in the usual telephone. However, each of the telephone numbers is equipped with a light which glows directly above the number being called. This makes it easy to transfer the call from one telephone to another. In some setups, there are additional numbered pushbuttons and these can also be used for call transfers.

PLE

An abbreviation for *public local exchange,* it is the type of exchange to which you are probably a subscriber and is the exchange operated by companies such as GTE, AT&T or possibly by one of the small independent telephone companies.

These exchanges are local. However, via a Toll Exchange, PLE subscribers can be connected via telephone to other Public Local Exchanges, either nationally or internationally. The lines used to interconnect toll exchanges are known as trunk lines or toll lines.

WATS

WATS is an abbreviation for *wide area telecommunications service,* an unlimited interstate or intrastate service. The service may be part time or full time, can be national or operate over a limited geographical area. A business may use a WATS number to permit customers to call without charge.

FTS

The *Federal Telecommunications System,* or FTS, is a direct-dialing telephone system used by Federal Government Agencies. It has a multitransmission capability and can handle voice, voice which has been encoded (scrambled), high-speed data, facsimile, and teletype.

SPRINT, MCI, COM, NETWORK ONE, TELTEC, WU

These are the names or abbreviations of companies that supply alternative long distance calling services on a national basis. There are also alternative long distance calling services that supply telephone communications within a state only.

ACCESS CODES

An access code is a number or group of numbers that let you connect to a line or service. Thus, numbers such as 1, 9, 800 are access codes. The access code for an alternative long distance line can be 7 digits. For the Bell system, the access code for a long distance line is simply the digit 1. Digit 9 may be required to access a local line, while 800 is used to access a long distance line to make a call for which you will not be charged. Note that the access code always precedes the number you want to call. In some instances it may be necessary to use two access codes.

In some cases dialing an access code eliminates the necessity for dialing an area code number. Thus, in dialing an 800 number it is not necessary to use an area code number. Just dial 800 followed by the 7-digit number you are calling.

POWER SOURCES

Electronic devices can be passive or active. A passive device is one that requires no source of power such as that supplied by batteries or the ac power line. An active device is one that must have electrical energy delivered to it before it can work.

A telephone is an active device and it gets its operating energy from the telephone lines. Your local telephone company delivers electrical energy to your home just as your local electrical power company does.

CUSTOM CALLING SERVICES

It is not generally known but you can get the assistance of your local telephone utility to obtain special help known as *Custom Calling Services*. These services are available in many areas but to individual line customers only. You can use the services with pushbutton or rotary dial telephones and there are no accessories required. Custom Calling Services are handled completely through the telephone utility's Central Office. The services are:

Call Waiting

This is a service that lets calls get through even when your telephone is busy. Generally, when you have a telephone call and are responding, someone trying to reach you cannot get through until your line is cleared. With this service, however, an urgent call can break into your conversation and give you the opportunity of deciding if you want to accept the interrupting call.

Call Waiting Service sounds a tone when you are on the phone and someone is trying to reach you. Simply press your receiver button to put the first caller on hold while you answer the second. You can even switch back and forth between the two. So you never miss important calls.

Call Forwarding

If you plan to be away from home you can arrange to have your calls transferred to where you can be reached. Known as call forwarding, it is a useful service if you expect to be away for a definite period of time at some place having telephone availability. The alternative to Call Forwarding is to use a telephone answering machine, such as those described in Chapter 7.

Call Forwarding is helpful to sales and professional people and others who rely on their home phones for appointments and information. It's a security measure, too. When you're away on a trip, transfer your calls to someone so your phone won't go unanswered. In some areas, calls can be forwarded to a long distance number with the long distance charges billed to your phone.

Three-Way Calling

With this service you can add a third party to a conversation and do so without the help of a telephone operator. Three-Way Calling Service is an easy way to set up your own conference calls. Just dial the first number, put that person on hold and dial the second. Now you're all on the same line. Two people with local numbers can talk to a third person long distance for the price of one long distance call.

Speed Calling

Speed calling is a way to dial telephone numbers that are frequently called, and could also include emergency numbers. The advantage of Speed Calling is that it lets you reach a desired telephone number in a fraction of the time it takes to dial, even using a touch button telephone. However, there are now telephones available, as described in Chapter 3, equipped with a memory or last number redial that let you handle speed calling without the assistance of your local telephone utility.

Speed Calling Service lets you make a phone call by dialing a one-or two-digit code. It is quicker, easier, and more accurate than dialing the full seven digit number (or more, if long distance). It is highly suitable for frequently called, emergency, and hard-to-remember numbers. It is especially helpful for children, and the elderly or handicapped. Speed Calling is available for 8 or 30 local and long distance numbers. You can make up your own personal list and change it as your needs change.

For all of the above listed services, it is necessary to get in touch with the service representative of your local telephone company. Since these are additional services you can expect to have your telephone bill increased accordingly.

Conference Calling

It is possible to have a conference telephonically, with as many as 10 telephones interconnected by the telephone utility. With conference calling the persons "attending" the conference can be at widely different locations, taking into consideration time changes. However, by using a speaker-type telephone (see Chapter 3) it is possible to have an in-company and intracompany telephone conference. Thus, groups can talk to groups, or an individual can talk to a number of persons simultaneously.

Data Service

Using a pushbutton telephone linked with a computer, it is possible to perform business functions, including billing, credit checking, inventory, travel reservations, banking, etc.

Data Phone

Using punched cards or magnetic tape, data can be transmitted, including charts, numbers, drawings, and photos.

Private Line Service

This is for the transmission of calls and data between two fixed points with no time limit.

Building Service

Your local telephone utility is available for consultation for home or business telephone wiring and will help in planning telephone layouts. There is no charge for this service.

INTEROFFICE TELEPHONES

It is possible to have telephones in offices intended for interoffice communications. These telephones are not equipped with outside lines. On-premises lines can be installed by a private contractor.

Interoffice telephones have a number of advantages. There is no charge by the local telephone utility. Since the telephones do not connect through the office PBX they do not tie up incoming telephone lines. There is greater telephone security since these telephones do not connect to outside telephone lines. And telephone conferences can be held without the need for assembling everyone in a single office. They have an advantage over intercoms since conversations are less likely to be overheard. Interoffice telephones can be equipped with extensions so two or more persons in the same office can use the in-office telephone system simultaneously.

SWITCHBOARDS

Calling from one in-home telephone to another in-home telephone involves switching only at the Central Office of the telephone utility.

In a business, however, a call often needs to be switched further, from the call receiving office telephone operator, to some person in the business oganization. These are handled by private branch exchanges or PBXs. At one time PBXs were cord plug-in types.

It is also possible to obtain a telephone that is a combined phone and call exchange, and these can be desktop and cordless types, equipped with as few as 12, or as many as 30, keys. Some permit in-office telephones to be used as intercoms, allow conference calling, or call transfer.

Console PBX

This is a desktop. switching system using a key-type board. Some are intended to handle incoming calls only. Outgoing calls need not go via the company's telephone operator and can be made directly. With some PBXs, all incoming and outgoing calls must be made through the office operator, with this operator supplying an outside line upon request.

TRANSMISSION OF THE TELEPHONE SIGNAL

After being converted into an equivalent electrical signal by the microphone in the telephone the signal moves through conductors (wires) to the Central Station. The voice in its electrical form can continue on to the called station via other wires, or it can be transmitted by radio, or sent through optical fibers. It can be transmitted to a satellite and from that satellite, using a transponder, returned to earth for further routing.

BELL OPERATING COMPANIES

The Bell operating companies throughout the U.S. include:

* Central and Pacific Telephone Companies
* Cincinnati Bell
* Illinois Bell
* Indiana Bell
* Michigan Bell
* Mountain Bell
* New England Telephone Company

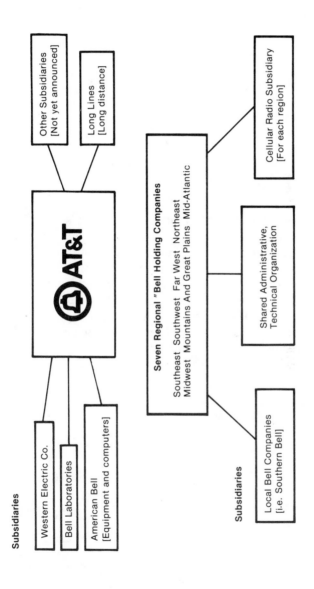

Fig. 1-19. AT & T and its subsidiaries; the seven regional Bell holding companies own the local operating companies.

- N.J. Bell
- N.Y. Telephone Co.
- Northeastern Bell
- Ohio Bell
- Pacific Northwest Bell
- Bell of Pennsylvania
- Diamond State Telephone Co.
- South Central Bell
- Southern Bell
- Pacific Telephone
- Southern New England Telephone Co.
- Southwestern Bell
- Wisconsin Telephone Co.

AT&T has a number of subsidiaries including Western Electric Co. (manufacturing); Bell Laboratories (research); American Bell (sales of telephone equipment); Long Lines (long distance communications), plus other subsidiaries to be announced.

There are various Bell holding companies including Southeast Bell, Southwest Bell, Far West Bell, Northeast Bell, Midwest Bell, Mountains and Great Plains Bell, and Mid-Atlantic Bell. These holding companies operate the local Bell Telephone Company utilities. When cellular radio starts functioning, it will be a subsidiary of the seven regional Bell holding companies. See Fig. 1-19.

CHAPTER 2

HOW TO SAVE ON TELEPHONE BILLS

There is no question the telephone has more than proved its worth and it is quite difficult to think there was a time when it did not exist. Still, every time you use the telephone, it is equivalent to making a purchase, but it is more like a credit card than buying for cash. As a result it is easy to overspend. But shopping for the best telephone rate is like shopping for any other commodity. It takes constant awareness that the end of the month will bring the inevitable bill. That bill is essential since telephones are essential. With just a little effort, that bill can be kept to a minimum and that does not mean eliminating necessary calls; it does not mean skimping on the number of phones and extensions in the home; and it does not mean a strict telephone budget. There are many ways, practical and sensible ways, of keeping your telephone bills within the limits of your budget.

The fact that your telephone bill represents computer output does not mean it is infallible. Computers may not make mistakes but people do, and it is people who supply computers with data input.

Considering the tremendous numbers of telephones in existence and the tremendous number of telephone calls that are made, it is not at all surprising there should be an occasional error in your monthly phone bill. It is always possible for someone to charge a long distance call to your number. It is always possible you may be charged for more calls than you are making. It is always possible for your phone to be used without your consent or knowledge. People, generally law abiding and honest, often have no compunction about making phone calls, particularly if they will not be presented with the bill. This amounts to theft, but it is not regarded as such.

If you keep a record of your calls, then checking your telephone bill should be easy, provided you understand what all the charges are for, and that includes charges for services in addition to those being made for calls.

One way of keeping track of telephone calls is to set up a telephone log, with all the members of your family agreeing to enter information about their calls. However, this is practically impossible to do. This information, plus a pocket calculator, will help you check your telephone bills quickly.

Don't hesitate to challenge any calls which *do not* appear in your telephone log, or (even if you don't keep such a log) which you don't remember. Usually local telephone utilities are quite agreeable about giving credit for unwarranted charges, particularly if you can back up your statements with a written record.

CHECK LIST FOR CUTTING TELEPHONE COSTS

There are a number of ways of reducing your telephone costs, and even though some of these may be small (such as for the single phone in a home) they can be substantial if considered on an annual basis. Even more so if a number of phones are used, or if you are operating a business.

1. Own your own telephone or telephones. Since 1977 it has been legal to buy your own telephones and to install them yourself. In doing so you make a double savings. You no longer need pay a monthly rental charge and you need not pay an installation fee.
2. You can buy your own telephone(s) from your local phone company or from a number of independent companies. Generally, the independents offer not only a larger variety, but at a more competitive price.
3. If you do buy your phones from your local utility, they may try to sell you a service contract at the same time. These contracts can add substantially to your annual phone costs and generally are not necessary. For telephones, you will find it more economical to be self-insured. If, over a period of one year, you do not need telephone servicing, then the service contract money you would have paid out will probably more than pay for the cost of future repairs, if any.

4. If you decide to buy your telephone from your local utility make sure you understand the purchase of the phone is for the complete unit, and not just the outside shell. In some instances, the utility will lease the phone mechanism itself and sell only the phone's external housing. This negates the entire idea of owning your own phone.

5. There are two basic types of telephones — the older style having a rotary mechanism for dialing calls and the newer pushbutton phones. Pushbutton phones are better for two reasons. They work much faster than rotary types. And it is possible that rotary phones will ultimately be phased out and become obsolete. If you have rotary phones and plan to replace them with pushbutton types, check with your telephone company to make sure the phone lines coming into your home or business can work with such phones.

   If your telephone lines are not geared for handling pushbutton phones, you can solve the problem by buying so-called *universal dialers*. These are pushbutton phones, but they do handle your calls more slowly than true pushbutton types.

6. If you have a number of telephones, such as in a small business for example, you may be paying a rental charge for the wiring that interconnects those phones. In some states, such as New York, for example, you can have your local telephone utility remove the wiring and you can install our own. In effect, the phone utility does not sell you the wiring — it just rents or leases it to you. Installing your own wiring is a one-time charge. If you are reluctant to do your own wiring, check with some of the independent companies that sell telephones and telephone accessories (such as wire, plugs, jacks, adapters, etc.). Some of these have an installation service or can make recommendations.

7. If you make a fairly large number of long distance calls, check with companies such as MCI and SP. Give their representative an opportunity to check your long distance calls by examining your old phone bills. They can then give you an estimate on the average monthly savings you can make.

   Be careful in your selection of long distance phone companies. The more reputable companies include Sprint, MCI, ITT, and Western Union. They may be listed in your local telephone directory.

8. Don't assume your telephone bill is correct. The arithmetic may be, but the charges may not be. If you run a small to moderately

sized business you may be interested in having an outside consulting firm examine your bills. One such company is Ronald Chernow Associates, Inc., Briarhill Bldg., Cornwall Court, P.O. Box 867, East Brunswick, NJ 08816. (201) 254-0100., If you have a large business you might consider using the services of National Utility Service, 44 Montgomery St., San Francisco, CA 94104. (415) 421-8800.

9. If you have a business, whether small or large, you will find employee abuse of the telephone quite common. Abuse in this case means using a company phone for personal long distance calling, unnecessary use of the phone, and demanding operator assisted calls instead of the less expensive direct dialing.

Telephone abuse is directly related to company policy. Some companies insist each employee enter calls in a log. A monthly review and a comparison of the log with phone bills soon pinpoints the existence of unneeded calls. Unless a company establishes some form of phone discipline, it soon learns that employees look on phone usage as a right.

Requiring a log can be a nuisance and it takes time. Another method is to set up a semirestricted phone system. With this arrangement long distance calls can be made only through one or two phones. You can also have semirestricted phones which can receive calls but which cannot be used for making them. Long distance calls can also be controlled if the business has a switchboard and a telephone operator.

It is also possible to set up a series of toll restrictors. These permit some long distance calls to be made, but not others. Thus, if in your business you must make frequent long distance calls to a certain geographical area, you can toll-restrict some or all of your phones for that area.

HOW TO SAVE ON LOCAL CALLS

Some people who shy away from making long distance calls disregard the cost of local calls based on the assumption they can make an unlimited number of such calls and talk for any length of time.

Unlimited Local Call Service

Whether it is possible to do so varies from one area to another. In some regions you can pay a flat fee for as many calls as you wish to make, provided these calls are all within a geographical region defined by your telephone company. This does not mean all telephone utilities automatically supply such a service or must do so. Either check your telephone directory or call your local utility to make sure.

Extended Area Service

Some telephone utilities offer what is known as *Extended Area Service.* This is the same as unlimited local call service, but since the geographical area covered is larger your phone bill will be higher. However, it may be more economical for you to subscribe to this service if you must make frequent calls within the larger area. One way is to keep a record of all the calls you make for a month or two. You can then determine if it is more economical to pay for Extended Area Service or to subscribe only to limited area local coverage.

Low Use Residence Plan

In some parts of some states there is available a *Low Use Residence Plan.* This plan may save you money if you make just a few local calls each month and these calls are usually brief.

Under this plan there is a limit to the number of calls you can make without additional charge. To learn more about this service, and if it is available, call or visit your local telephone utility. Even if the plan is being used in your area, you must still check with the business office of your telephone company.

Once you become a member of the Low Use Residence Plan follow it thoroughly, even if it means keeping a record of all your calls and their time duration. Generally, the Low Use Residence Plan is best suited for those homes equipped with just a single telephone. With one or more extension phones and more than one user, it becomes difficult to keep track of all the calls being made, to say nothing of the problems involved in getting cooperation from all those using the phones. Again, those who do not pay for the calls are the ones for whom frequent calling and lengthy calling is most attractive.

Telephone Planning

If you subscribe to the Low Use Residence Plan and if saving money on telephone bills is essential, a little planning will be helpful.

List the things you want to talk about before making the call. Then cross these items off as you go along. You'll cover more in less time. And you won't have to call back because you forgot something.

Local Measured Service

The various public telephone companies have come up with a variety of billing plans and so it is sometimes difficult to keep up with them. One of those that has been proposed is referred to as *LMS* or *Local Measured Service*. With this plan the charge per local call is lower, but there is a limit on the total number of such calls that can be made. After the local call quota has been reached, local calls would be charged on a per-minute basis, just as though the local calls were long distance. But even though above-quota local calls would be on a per-minute basis, there may be a 50% discount for calls placed during certain times (just like long distance calls) and there may also be a rate package for those who consistently do not reach the local call quota.

If you don't know whether to use the Low Use Residence Plan or the Local Measured Service Plan, you can try either for a month or two (assuming these are available) and then cancel if the savings do not come up to your expectations.

DISCONNECTING YOUR LOCAL UTILITY TELEPHONE

If you plan to move from a house or apartment that contains one or more telephones rented from your local utility it is essential that you notify your local telephone company and supply them with the termination date, otherwise you could be charged for calls made after you move.

Disconnecting a telephone is easy and involves nothing more than removing the plug, whether modular or standard. If you return the telephone to your local utility you will receive a credit, usually $5, on your final telephone bill.

Suspended Rates

If you are planning a vacation or a trip for several weeks, or if you operate a business that closes completely for about a month, you can get a reduced rate during the time of your absence. Generally, the local utility has a minimum time requirement such as two weeks for a home telephone, four weeks for a business phone. While you do save on temporary discontinuance of your phone (or phones) there may be a charge for the resumption of service.

If you have a business that has more than one telephone number, you could keep one line active and the other inactive for the time of your absence. Using a telephone answering machine on the remaining active line would inform callers that your place of business is closed temporarily. You might also check with your local telephone utility to learn if calls made to your inactive line could be transferred to the active line so that callers could receive your telephone answering-machine taped message.

However, if you plan to have all your business telephones disconnected, you might also arrange to have all your incoming business calls transferred to your home or to the home of someone who has agreed to take all your calls for you.

DIALING A WRONG NUMBER

If you dialed a wrong number and the call was completed, but not to your party, hang up and call the operator at once. You will receive credit for the call. If the call was a local one and you have an unlimited number of calls available as part of the service, then asking for credit is unnecessary. But for limited local calling privileges and certainly for long distance calls you should ask for a credit.

LINE INTERRUPTION

You can ask for operator assistance to interrupt a busy line in the event of an emergency. There is a nominal charge for this service, usually about 30¢ for each time you request such an interruption.

CALLING COLLECT

You can call any other telephone number collect but in such instances someone at the called number must accept the call and agree to pay for it. If you are calling person-to-person, the individual called must agree to pay the charges. If you are calling station-to-station, someone at the receiving end, usually a telephone operator, must be willing to accept the call and pay for the charges.

OPERATOR ASSISTED CALLS

These are the calls that carry the highest rates. Operator assisted calls include person-to-person calls, calls made using a coin-operated telephone, collect calls, calls using a telephone credit card, calls that are to be charged to some other number, charge calls, and calls you make as a guest in a hotel or motel. Hotels and motels may add their own service charges to those set up by the telephone utility.

All operator assisted calls have a three-minute minimum, unlike dialing direct with its one-minute minimum. Hence you do not make any savings with an operator assisted call by keeping the calling time as much below three minutes as you can. With operator assisted calls, the charge beyond three minutes is the same as direct dialing, using day, evening or nighttime rates, depending on which is applicable.

Person-to-Person Calling

All person-to-person calls are operator assisted and so are more expensive than direct dialing. This type of calling is somewhat of a gamble. If the person you are calling is in, you pay the operator assisted rate. However the charge does not begin until the called person answers the telephone. If that person isn't immediately available the waiting time is no cost to you. Further, there is no charge for the call if you do not make connections with the person you requested.

If you work in an office and receive a telephone call memo asking you to call an operator by number (such as "please call operator 5") and you do so, the charge will be that of an operator assisted call. It will be less expensive to use direct dialing, making the call yourself without the help of the operator. This assumes you have the name and telephone number of the person who called you or that you have some way of getting this information.

Directory Assistance

Whether you use your telephone(s) at home or in business, or both, your local utility supplies you with a telephone directory. It is often much easier, though, to call on an operator for assistance. However, the telephone company can charge you after you have made six requests for directory assistance (within your monthly billing period). The charge is 15¢ per assistance request. If you must use this telephone service try to ask for information about two numbers at the same time. With this method the two numbers furnished will count only as one, and so you will be able to obtain a maximum of 12 before charges for directory assistance will begin.

If you know a telephone number has been changed but you do not have it available, call the old number anyway. An operator or recording will come on the line and will supply you with the new number. In this way you will avoid a directory assistance charge.

UNDERSTANDING YOUR TELEPHONE BILL

Because there are now a number of independently operating telephone companies telephone bills can vary in their details from one area to the next. If you don't understand your phone bill, a telephone call to the business office of your telephone company should supply you with the information you need. A visit to that office is even better.

One of the difficulties with telephone bills is that charges are often grouped. Thus, all bills have a monthly service charge. What isn't generally known is what this monthly service charge includes. The basic monthly charge isn't just one, but is a summation of three charges. The first of these is a charge for a telephone supplied by your local telephone utility. If you own your own telephone this should not be a part of the basic monthly charge. The problem is that if this charge comes under the heading of "basic monthly service charge" you do not know whether or not you are being charged for telephone "rental." A query to the business office should clear this for you.

Another part of the basic monthly charge is the inside wire charge for connecting telephones to the exchange access line. And, finally, the basic monthly service charge includes a charge for the exchange access line from the Central Office to your home or business.

Your bill may also have an on-premises wire charge. This cost is made up of two parts: a wire-investment charge and a wire-maintenance charge. If you have installed your own on-premises wiring it would be advisable to challenge this charge. The wire maintenance charge means the local telephone utility is charging you for the cost of maintaining the wiring, and this is correct only if the telephone utility has supplied your on-premises wiring. If they did so originally, but you subsequently installed your own on-premises wiring you should no longer bear this charge.

The charge for on-premises wiring is a charge that lets the telephone utility recover its investment in premises wire. Again, if you have your own on-premises wiring you should be free of this monthly charge.

If you have more than one telephone, there could also be a wire charge for each. If you wired in your own telephones this charge should not be included in your telephone bills.

CHALLENGING THE LOCAL TELEPHONE UTILITY

If you have any questions about your telephone bill, or what each item represents, get in touch with the business office of your telephone company. The telephone company has every right to be paid for their services, but since you are paying the bills you also have the right to know exactly what you are paying for. Telephone companies are operated by people and people do make mistakes. You will find telephone companies to be courteous and helpful and their personnel are trained to give you the information you need, and to do so cheerfully.

BUNDLING

At one time the telephone bill you received may have been bundled, that is, all charges were incorporated and listed as a single item. Today telephone bills must be categorized, broken down into separate units such as rental, wiring, service, taxes, local charges, and long distance charges.

FCC REQUIREMENTS FOR TELEPHONE OWNERSHIP AND USE

When you interface your telephone(s) with a telephone company line, that is, when you connect your phones to the line, you are, in effect, entering someone else's property. This does not mean your local utility can arbitrarily deny you the use of their lines, but to be allowed to do so you must comply with FCC rules. For prospective owners of telephones and telephone equipment this means notifying the telephone company of the registration number and the ringer equivalence number. Fig. 2-1 shows what this could look like on your new units or on the shipping container.

The FCC Registration Number is a combination of 14 digits and letters. The numbers use the decimal system and range from 0 through 9. The first six will be a combination of numbers and letters and will be AA499C. Following this will be five digits and then three letters.

At the same time you should notify the telephone utility of the telephone number of the line to which you are connecting the equipment. This might not be necessary if you have just a single telephone number, but will be required if you have more than one of these. You should also notify the phone company that the jack required for your device is USOC RJ11C.

USING THE TELEPHONE DIRECTORY

There is often a difference between in-state and out-of-state telephone regulations. If you are in an out-of-town location and expect to make a number of calls, both local and long distance, it could be to your advantage to consult the introductory white pages of the local telephone directory. This will give you some idea in advance of what your telephone expenses could be.

OWNING YOUR OWN TELEPHONE

You can reduce your monthly telephone bill by a few dollars by owning and installing your own telephone(s). It is perfectly legal for you to do so, assuming the phones you buy are FCC type approved.

In some states, local telephone utilities are expediting the process of individual ownership of telephones by offering their subscribers a purchase option. This makes sense for both parties — the subscriber and the telephone companies. The subscriber, in buying the tele-

---

> **Complies with Part 68, FCC Rules**
> **FCC Registration Number**
> **AA499C-00000-XX-X**
> **Ringer Equivalence 0.0X**

Fig. 2-1. Type acceptance of telephone equipment. The zeros shown will be numbers 0 through 9. The letters X will be the letters of the alphabet from A through Z.

phone, avoids payment of a monthly rental charge. The telephone company manages to dispose of a lot of telephones that are used.

However, the subscriber has an additional, attractive option. He can now shop in the open market for a telephone, and now has a tremendous shopping variety offering features and styling previously not available. Even though the "buy your own phone" approach by telephone utilities seems attractive, it is prudent to do some shopping first.

To encourage the purchase of their telephones, some utilities are offering the units on a sort of installment plan. Instead of paying for the telephone as a single cash purchase, subscribers pay a slightly higher monthly charge, ranging from about 60¢ to $1.00. When this surcharge reaches the price of the telephone the unit belongs to the subscriber.

Telephones rented from local telephone utilities have always had the advantage of a lifetime free warranty. With the new ownership arrangement that warranty drops to 30 to 90 days, depending on the local utility. For repairs beyond those dates it is necessary to return the phone to American Bell.

INSTALLING YOUR OWN PREMISES WIRING

Most premises telephone wiring is that installed by the local telephone company and you pay a monthly service charge for it. You can eliminate this charge by installing your own wiring. Some stores that sell telephones also have a wiring service or can recommend someone who can do the job for you.

THE DEFECTIVE TELEPHONE

When you buy your own telephone it is possible it may cause some problems on the telephone line. In that case the telephone company has the right to discontinue your service temporarily until the problem is cleared. You may have purchased a defective phone, or the telephone may have worked well for a time and then developed a problem.

The telephone utility does not have any responsiblity for the telephone you purchased, but you have a responsibility for making certain you do not interface defective equipment with their lines.

If the telephone is still under warranty, you can return it for repairs, generally without a charge for parts or labor, depending on the wording of the warranty. If out of warranty, you will need to pay all repair charges, including materials and time, or else replace the phone.

In the event your telephone does cause telephone line problems your telephone utility will notify you. Disconnecting the phone is simple and is just a matter of removing the modular plug from its wall jack. The company from whom you bought your phone may rent you a telephone, pending the repair of your unit, or may loan you one without charge if your phone is still under warranty. Every store, whether a dedicated telephone store or some other kind, has its own policy toward its customers.

If the telephone is a one-piece unit, and if the defect is serious enough, you may need to replace the entire phone. If it is a two-piece unit you may need to replace only the handset or the base containing the dialing pad. But even with two-piece units this isn't always possible.

Noisy Telephone Lines

If your telephone has been working well and then suddenly becomes noisy, or if the persons you call or who call you sound very weak, then it is possible there may be some trouble with the telephone line. Sometimes the connection that is made for you is a poor one and a new connection should be tried.

Sometimes a telephone line will be extremely noisy or you may not be able to hear or understand the person you have called. In that case, both parties should hang up. Call the operator by dialing 0 and explain what has happened. If you made the call an adjustment should appear on your phone bill, probably in the form of a credit. Only the caller can receive a credit; not the called person.

Dialing Problems

It is sometimes possible to have trouble when making a call. If that happens hang up the receiver and try again. You may not have pressed a pushbutton completely, you may have pressed two pushbuttons simultaneously or you may have omitted pressing a pushbutton. Hang up the phone, wait a moment and then try again. Be conscious of the fact that you are pushing buttons and reduce the speed with which you are handling them.

THE TELEPHONE TAX

There is a Federal Excise tax on telephone bills. The provisions of the Tax Equity and Fiscal Responsibility Act of 1982 established a 3% tax on telephone services, starting January 1, 1983. This tax rate will be effective through 1985. Under present federal law, the tax will be eliminated January 1, 1986.

COST OF LOCAL SERVICE

Telephone utilities have long maintained that local service calls are, in effect, subsidized by the more profitable long distance lines. Consequently, when the industry was deregulated (AT&T has divested itself of its 22 telephone subsidiaries throughout the U.S.) and became more competitive, claims were made by telephone utilities for an increase in local calling rates. That is why finding every avenue of approach and every technique for lowering your telephone bill is so important. The cost of local telephone service will continue to rise.

TELEPHONE DEPOSIT

When you request telephone service for an apartment, house or business, you may be required by your local telephone utility to supply a deposit. Whether you will be asked to do so or not depends on your "credit" ability. You may be requested to supply some information about your financial status — whether you own your own home, if you are employed, and a bank reference. You may also be able to get phone service if you can supply the name of a guarantor — an individual or a company that will guarantee payment, assuming, of course, their ability to establish a line of credit.

The amount of the deposit will vary, depending on the number of local and long distance calls you estimate you will make and the credit line you have established. It changes from one public telephone company to another, based on the credit policy they have established.

Interest is paid by the local telephone company on all deposits, but there is no established rate for all companies. You can ask for a return of your deposit at the end of a certain period (generally about two years or so) if you have paid all your telephone bills on time. If your telephone service is disconnected, your deposit plus accrued interest is applied to your telephone bill, with any surplus returned to you.

ADVANCE PAYMENT

If you have just moved into an area and want immediate use of a telephone, and if installation is involved, you may be asked to make an advance payment of the installation charges plus a payment for one month's local service, pending establishment of your credit. If your credit is satisfactory, this advance may be subtracted from your first month's telephone bill.

Whenever you make a payment to your local telephone company, whether an advance deposit or advance payment, be sure to ask for a receipt. It is preferable not to pay cash but to make payment by check or credit card. In this way you will have two pieces of documentation about your payment — the receipt and the cancelled check or bill from your credit card company. You will need some proof to show the telephone company at the time you ask for the return of your deposit. The reason for having a telephone receipt plus your cancelled check or credit company bill is there is always the possibility you might lose one. Having the other is still proof of payment.

CREDIT CONFIDENTIALITY

You need not be concerned that credit information you have supplied to your local telephone company will be made available to other individuals or companies. Telephone companies will not honor requests for credit information about their subscribers.

TWO-PARTY LINE VS PRIVATE LINE

Having a two-party line means you share your telephone line with some other individual. The advantage is that you can save approximately $2.00 a month or a little more on your telephone bill. The disadvantage is exactly what sharing implies — you can only use your phone when the other party isn't using it. They can also listen in on your conversations (and vice versa).

TELEPHONE COMPANY SERVICE CHARGES

When you ask your telephone company to perform a service for you then you can expect to receive a bill for that service (or services). Although your telephone book lists the costs of the various services, it is prudent to call the business office of your telephone company to check since service charges can and do change.

Service Order Charge

In terms of payment this is one of the largest service charges, and is in excess of $30. This is the service charge made when you install or make changes in your telephone service.

Central Office Connection Charge

This charge is approximately $20 and is made for connecting your line to the Central Office switching system.

Premises Visit Charge

If it is necessary for a telephone company technician to visit your premises there is a premises visit charge, and is approximately $6.00.

Premises Wiring Charge

If you want the telephone wiring changed or moved, there is a charge of about $13.50. You do not pay this charge if you install your own on-premises wiring.

Jack Installation Charge

If you want a telephone jack installed it will cost you about $7.00 for each. You do not pay this charge if you buy and install your own jacks.

Telephone Handling Charge

There is a charge of $4.50 for each telephone your company connects for you. If, for example, you bring along your own phones from a former residence, you can plug them in yourself or you can ask your phone company to do it for you. Plugging a telephone into a jack is as easy as connecting a toaster or radio to an ac outlet. You can make a substantial saving if you do it yourself, or possibly get a friend to do it for you. Plugging in your own telephone(s) is perfectly legal.

Overall Charges

Telephone company charges for the various services listed in this chapter can be different from one geographical area to the next. Further, these charges can vary from time to time, generally in an upward direction. Consequently, the rates indicated on this and preceding pages should be regarded as approximations only.

TAKE AND SAVE PLAN

If you plan to move but will still be using the services of your present telephone company and if you have been renting your telephones from that company, check with them to determine what savings you can make. You may be permitted to take your phones along and connect them in your new location.

OPTIONAL SERVICES

At any time you see or hear of an offer for optional services you can be sure these will increase the size of your telephone bill. You may want these services, you may find them convenient, but they never come under the "freebie" heading.

Nonlisted Semiprivate Number

You may not want your number, name and address in the phone directory so as to avoid unnecessary calls. You can have a nonlisted semiprivate number. Your telephone number, however, will still be available to others through Directory Service Assistance.

Private Unlisted Telephone Number

You can have a completely private telephone number, not listed in either the telephone book or available through Directory Assistance. The only persons who will be able to call you will be the ones to whom you have given your telephone number.

Additional Listings

When you subscribe to telephone service you are automatically entitled to a listing in the white pages of the local telephone directory. You can have an additional listing if you wish, but you will have to pay for it. An additional listing might be necessary, for example, if two persons having different names share the same household and the same telephone.

CUSTOM CALLING SERVICES

Your local telephone company may have an option known as *Custom Calling Services,* and, like any other options, these will increase your monthly telephone bill. These include:

Call Waiting

When you answer your telephone there is always the possibility someone will call you while you are talking. The second party will hear a busy signal. If you want to avoid this possibility you can subscribe to a *Call Waiting service.* While you are talking on the phone you will hear a special tone to alert you that another call is waiting. At that time you can depress the switchhook and "hold" the first call while answering the second.

The alternative is to have not one, but two, telephone lines — that is, to have two telephone numbers. The Call Waiting option costs less. However, having a second phone with an independent phone line means two persons can receive or make outside calls independently.

Call Forwarding

With this option you can transfer your incoming calls to any telephone number.

Speed Calling

This option lets you program your telephone to dial numbers you call often by simply depressing one or possibly two keys. Since this becomes a regular monthly increase in your telephone bill, a more economical alternative is to buy a telephone having a memory. In this way you can save twice. You can save the telephone monthly rental charge and also the charge for the option of Speed Calling.

Conference Calling

There are two ways of talking to two or more persons at different locations and having different telephone numbers. You can call one number first, and then, upon completion of the call, telephone the second number. An alternative method is to use the *Conference Calling* option. With this option you can add a third party to your telephone call. If you do a large amount of conference calling, you could make a comparison of the charges (and time) involved in making the additional telephone calls vs the three-way calling option.

TDD

TDD is an abbreviation for *Telecommunications Device for the Deaf*. For someone who is deaf, or is otherwise disabled, your local telephone company has special equipment available, described in Chapter 3.

TDD Long Distance Rates

Disabled customers who use a TDD to communicate over phone lines are eligible for reduced rates on long distance calls between states dialed direct from a certified resident phone. Many state regulatory commissions have approved similar reductions for long distance calls dialed within the state. Check your local Bell company to obtain the necessary certification forms to apply for the lower rates.

TDD Operator Services

TDD users who want to reach a telephone operator can do so by dialing 800-855-1155 (you must dial "1" first in some communities). The TDD operator can obtain numbers for you from Directory Assistance and can help you place long distance calls (such as credit card calls, person-to-person calls, collect calls, and others). Long distance calls placed through the TDD operator cost more than long distance calls you dial yourself. The TDD operator can also give credit for misdialed calls and can assist in completing calls with which you are having difficulty. The TDD operator can also report phone repair problems.

TDD Directory Listings

In most Bell companies, TDD customers have the option of listing their phone numbers in the telephone directory (or just in Directory Assistance records) with the words "TDD only" or "TDD & voice" after their names. TDD customers may also omit their addressed from their directory listings.

CREDIT FOR LOSS OF SERVICE

If your telephone is out of service for 24 hours or more and you have reported you have been unable to receive or make phone calls, you may be entitled to a credit for loss of service. However, it is questionable if you will get such a credit if the fault is due to an inoperative telephone which you have purchased and connected.

PERSONAL DIRECTORY

Because of its size, some people are reluctant to use the telephone directory. There are several ways of making this chore a bit easier. One method is to underline numbers you call frequently. Another is to set up your own little telephone directory.

EXTRA LISTINGS

When you become a subscriber to a local telephone utility you are entitled to one listing in the white pages of the telephone directory. However, if there are people with different names in your home you might consider the advantage of having more than one listing. There is a charge for the use of an extra listing (or listings) other than the original one.

Additional listings aren't free just because you have more than one telephone in your home. You can get additional listings by paying for them or by having more than one telephone number.

INTERSTATE VS INTRASTATE CALLING

The fact that a call is long distance does not mean it is just any call made to another state. Long distance charges are as applicable within a state as they are to outside states. In either instance, the lowest charge will be in effect if you do not require dialing on the part of the operator, that is, if you do all the dialing yourself. Note the rate is not affected by the kind of telephone you use, whether it is an inexpensive rotary dial or a more costly pushbutton, whether it is a simple or highly decorative unit. Long distance charges are not affected if you use more than one telephone, that is, if you use your main telephone and one or more extensions.

There is a myth to the effect that the shorter the distance the smaller the cost. As an example that this is just not so, consider that if you call from Pompano, Florida to Jacksonville, Florida, a distance of 326 miles, the charge for direct dialing a one-minute call is 63¢ while the charge, also for a direct dialing one-minute call from the same starting point, Pompano, Florida, to Boston, Massachusetts, a distance of about 1500 miles, the charge is 66¢. Interestingly the cost of a call from Pompano to Seattle, Washington, a distance of 3300 miles, is exactly the same.

RECEIVE ONLY TELEPHONES

A receive only telephone is one that does not contain a ringing circuit. As its name implies, it can be used for receiving calls but not for making them and so such phones are not equipped with dialing pads or rotary dials. They are excellent for use as extension phones as indicated earlier in Chapter 1.

Having a receive only telephone can be inconvenient, but consider that in the average home only one telephone can be used at a time for making outgoing calls. With one or more receive only telephones more than one person in the home can listen, at the same time, to an incoming phone call. There is no charge by your local telephone company for the use of receive only phones.

RATE COMPARISONS

The rate chart shown in Chart 2-1 is for the purpose of making a comparison of long-distance calling rates between selected cities. All the prices shown are for 20-minute calls using direct dialing. The time period is any time, all day Saturday, and Sunday until 5 pm. The same rates apply any night after 11 pm and until 8 am. These rates are not applicable for Alaska or Hawaii.

To find the rate between any two of the cities listed, select the two cities and trace a horizontal line from one and a vertical line from the other. The intersection point is the rate.

Thus, to determine the cost of a 20-minute telephone call for direct dialing during the time periods mentioned earlier, locate New York in the chart and move out horizontally. Locate Cleveland in the chart and move down. The intersection point is $3.45. This will be the cost of this call under the conditions indicated.

LONG DISTANCE CALLING

An easy way to save on long distance phone bills is to eliminate such calls during the so-called peak period — weekdays from 8 am to 5 pm. This peak period is the time the telephone lines are most heavily loaded.

Discounts are offered by your local utility to encourage you to make your long distance calls in the evening, at night, or over the weekend. Lower rates are also available on holidays such as New Years Day, July 4, Labor Day, Thanksgiving Day, and Christmas.

Maximum Rate

The maximum rate is the full day rate and is the most expensive for the first minute of the call. Each additional minute is approx-

Chart 2-1. Comparison of Long-Distance Rates for 20 Minutes of Direct Dialing at Selected Times

Cities (diagonal headers, top to bottom): ATLANTA, BOSTON, CHICAGO, CLEVELAND, DALLAS, DENVER, DETROIT, INDIANAPOLIS, MEMPHIS, MILWAUKEE, NEW HAVEN, NEW YORK, OMAHA, PHILADELPHIA, SAN FRANCISCO, SEATTLE

															3.64
														3.55	3.55
													3.45	3.55	3.55
												3.64	3.55	3.64	3.55
											3.55	3.64	3.64	3.64	3.64
										3.64	3.64	3.07	3.22	3.55	3.55
									3.22	3.64	3.55	3.22	3.22	3.55	3.45
								3.45	3.55	3.55	3.45	3.55	3.55	3.64	3.45
							3.55	3.22	3.22	3.55	3.55	3.45	3.07	3.55	3.55
						3.55	3.64	3.55	3.55	3.64	3.64	3.55	3.55	3.07	3.55
					3.07	3.55	3.64	3.55	3.55	3.64	3.64	3.45	3.55	3.22	3.55
				3.64	3.64	3.45	3.55	3.55	3.55	3.55	3.55	3.45	3.55	3.64	3.55
			3.64	3.07	3.22	3.55	3.55	3.55	3.55	3.64	3.64	3.45	3.55	3.22	3.55
		4.06	3.64	4.06	4.06	3.64	3.64	4.06	4.06	3.64	3.64	4.06	3.64	4.06	4.06
	3.55	4.06	3.64	4.06	4.06	3.64	3.64	3.64	4.06	3.64	3.64	4.06	3.64	4.06	4.06

imately 31% off the first minute rate. Thus, a long distance call could be 64¢ for the first minute and 44¢ for every minute thereafter.

Evening Rate

The evening rate is from 5 pm to 11 pm weekdays, Monday through Friday inclusive, and also on Sunday, at these same hours. The evening rate offers a 35% discount from the weekday full rate. Thus, if the full rate is 64¢ for the first minute, the evening rate is 41¢ for the first minute. Each additional minute of telephoning using the evening rate results in a discount of about 28% from the basic evening rate.

Night Rate

The night rate is the lowest of all rates and is effective every night, Monday to Sunday, inclusive, from 11 pm to 8 am. On Sunday, however, the economical night rate is in force only until 5 pm, at which time the evening rate takes over. The night rate is approximately 60% lower than the weekday full rate. As is the case with the full day rate and the evening rate, the first minute of the night rate is the most costly. Thus, the first minute of a night rate could be 25¢ and each succeeding minute 18¢, a discount of 28 percent. The most economical telephone rates for long distance calling is at night and the weekend.

ECONOMICAL CONVERSATION

Another common myth is that all long distance calls have a three minute minimum. This is true only for operator assisted calls, and, as mentioned earlier, direct long distance calling has a minimum of 1 minute. To keep calling time to a minimum you might make notes of what you want to discuss, keeping social amenities to an absolute minimum. There are also telephones available which will let you time your calls. These are described in Chapter 3.

ALTERNATIVE LONG DISTANCE TELEPHONE SERVICES

There are now at least six companies offering long distance telephone calling service in direct competition with AT&T. But since these companies are also competitors, the rates they charge are different. And so, if you are interested in saving money on long distance calling, it is essential to shop for long distance services much as you would shop for any commodity.

AT&T does not have a minimum monthly charge, as shown in Table 2-1. However, its charge per long distance telephone call is larger than those of any of its competitors. Western Union and Teltec have basic monthly charges that are minimum fees, but telephone calls can be charged against this fee. However, against this you should consider that their charge per long distance call is greater than those of the other services with the exception of AT&T and Network One. Network One has a monthly charge of $5.00, whether you make any long distance calls or not. It also has the highest telephone calling rate, with the exception of AT&T. However, it does offer nationwide service, something some of the other companies do not.

The charges shown in Table 2-1 indicate the cost of a 10-minute telephone call between Florida, New York, Indianapolis, and Los Angeles. What your costs will be depends on where you are located, the location of the city you are calling, and the length of time of the call. Except for those companies offering nationwide service, such as AT&T, Network One, and Teltec, the competitive long distance calling companies serve only between 110 and 226 cities and/or metropolitan areas. However, these companies maintain they are expanding their services. Table 2-1 is included only as an example of comparative costs. Costs may be entirely different in your particular area.

Advantages and Disadvantages of Alternative Long Distance Calling

The only reason anyone should want to use the services of an alternative long distance calling service is to save money. Making a telephone call using an alternative service can be frustrating. You must first dial a seven-digit access code. This is followed by the telephone number you are calling. And since this will be a long distance call it will consist of the area code and the telephone number, a total of 10 digits. This must then be followed by your subscriber number, consisting of 7 digits. That makes 24 digits in all. You will need a lot of patience to put through your call, particularly if the called number is busy or you are using a rotary dial telephone. Calling long distance using alternative long distance calling can be simplified by using a telephone with a memory capability.

The first number you dial with alternative phone service is an ac-

Table 2-1. Comparative Analysis of Costs of Alternative Long Distance Calling Systems. Data Supplied Is Sometimes Contradictory.

	MINIMUM MONTHLY FEE	CITIES SERVED	COST OF 10-MINUTE CALL BETWEEN 5 AND 11 P.M. FROM FLORIDA		
			TO NEW YORK	TO INDIANAPOLIS	TO LOS ANGELES
AT&T	—	NATIONWIDE	$2.81	$2.81	$3.14
CITY-CALL/ITT	$5	110	1.51	1.50	1.85
WESTERN UNION	10*	226	1.70	1.70	1.81
SPRINT/S. PAC.	5	166	N/A	N/A	N/A
MCI COM.	5/10★	200	1.63	1.63	1.83
NETWORK ONE	5	NATIONWIDE	2.50	2.50	2.70
TELTEC	11.11*	NATIONWIDE	1.90	1.90	1.90

* Calls billed against this fee, not in addition to.
★ $5 evening and night only; $10 for 24-hour service.

OUT-OF-STATE
(3 minute call, 8 AM-5 PM)

	Atlanta	NYC	Chicago	Portland Maine
Network 1	1.05	1.14	1.14	1.14
Bell	1.48	1.52	1.52	1.52
				NO SERVICE
MCI	1.11	1.14	1.14	you must
Sprint	1.11	1.14	1.14	go through Bell

INTRASTATE
Calls from Miami—3 Minutes

	Broward	Boca	West Palm Beach
Network 1	.48	.72	.84
Bell	.78	1.03	1.04
MCI NOT AVAILABLE			
Sprint NOT AVAILABLE		.57	

cess number, a seven-digit number that lets you get to or reach alternative phone service lines. But sometimes this access line is busy and so your long distance call is delayed to that extent.

With AT&T you do not need to use an access code and you can dial your long distance call directly or have your phone memory do it for you. The service is nationwide and you know you can make the call. But with alternative phone service you must first check a chart to learn if the long distance number you want to reach is accessible.

Table 2-2. Rates for Overseas Countries You Can Dial.

Region	Rate Levels	First minute	Additional minute	Hours
UNITED KINGDOM/IRELAND	Standard	$2.08	$1.26	7 am-1 pm
	Discount	1.56	.95	1 pm-6 pm
	Economy	1.25	.76	6 pm-7 am
EUROPE	Standard	2.37	1.33	7 am-1 pm
	Discount	1.78	1.00	1 pm-6 pm
	Economy	1.42	.80	6 pm-7 am
PACIFIC	Standard	4.22	1.58	5 pm-11 pm
	Discount	3.17	1.19	10 am-5 pm
	Economy	2.53	.95	11 pm-10 am
CARIBBEAN/ ATLANTIC	Standard	1.68	1.13	4 pm-10 pm
	Discount	1.26	.85	7 am-4 pm
	Economy	1.01	.68	10 pm-7 am
SOUTH AMERICA	Standard	2.77	1.18	7 am-1 pm
	Discount	2.08	.89	1 pm-10 pm
	Economy	1.66	.71	10 pm-7 am
NEAR EAST	Standard	3.68	1.33	8 am-3 pm
	Discount	2.76	1.00	9 pm-8 am
	Economy	2.21	.80	3 pm-9 pm
CENTRAL AMERICA	Standard	2.62	1.13	5 pm-11 pm
	Discount	1.97	.85	8 am-5 pm
	Economy	1.57	.68	11 pm-8 am
AFRICA	Standard	2.89	1.48	6 am-12 Noon
	Discount	2.17	1.11	12 Noon-5 pm
	Economy	1.73	.89	5 pm-6 am
INDIAN OCEAN	Standard	5.22	2.17	6 pm-1 am
	Discount	3.92	1.63	1 am-11 am
	Economy	3.13	1.30	11 am-6 pm

For countries that are not dialable, there's a 3-minute minimum and rates are somewhat higher.
Different rate schedules apply to Canada and Mexico. Check your local operator.
Federal excise tax is added on all calls billed in the United States.

This means you may find the number you want to call is not available with the alternative service and you must still use the AT&T system.

Alternative Long Distance Calling Prices

If you plan to use alternative long distance telephone service, consider that the service may offer a choice of rates. As an example, MCI has a basic charge of $10.00 a month if you want 24-hour service.

This drops to $5.00 a month if you agree to confine your long distance calling during evenings and weekends. Teltec does have nationwide service and it charges a basic rate of $11.11. However, if you restrict your telephone calls to 30 major metropolitan areas the rate drops to $8.88 a month.

THE 800 NUMBER SYSTEM

An 800 number (mentioned earlier in Chapter 1) can be used without charge to you, assuming the person or company you are calling has such a number. An 800 number is free only in the sense that you do not pay for the call, but is paid for by the person you are calling. These numbers are invariably used by a business to encourage customers to make calls. If you are planning to call a company it is to your advantage to learn if an 800 number is available.

There are a number of ways of obtaining 800 numbers and you will find it worthwhile to keep a record of those 800 numbers you must use from time to time. Companies sometimes list their 800 numbers in advertisements, in their promotional literature, or it may appear on their letterheads. During a telephone call you can also ask if an 800 number is available.

You can also learn if a company has an 800 number by dialing 800-555-1212. This is a no-charge telephone call. All you need to do is to supply the operator with the name of the individual or company you are trying to reach. Some publishers produce directories of toll-free 800 numbers and you may find these at your local bookstore. One such publication is the *Toll-Free Digest,* Box 800, Claverack, NY 12513. Their book, which contains 25,000 toll-free numbers costs $6.50 plus 50¢ for postage and handling. Warner Communications also publishes a Toll-Free Digest containing almost 15,000 listings and selling for $2.50.

Another publication along these lines is *The National Directory of Toll-Free Phone Numbers,* sold for $8.95 plus $1.50 for postage and handling, by Celebrity Publishing, 6 Doe Drive, Suffern, NY 10901. This publication lists approximately 20,000 numbers and contains 535 pages.

Still another 800-number directory, designed for travelers, is the *Toll-Free Travel-Vacation Phone Directory.* This book has 378 pages and costs $6.95. Another book for those who like to do their shopping at home is the *Toll-Free Shop at Home Phone Directory,* with 272 pages of toll-free numbers of catalog houses. Its cost is $6.95 plus

$1.50 for postage and handling. These books about toll-free numbers are available in bookstores, such as Walden Books or at Bell Phone Center stores.

THE 800 NUMBER PROBLEM

The problem with calling an 800 number is that such numbers have become increasingly popular, and, as a result, often results in a busy signal. You may get through only to be put on hold. It is true you are not being charged for the time you are being kept on hold, but it often means a long wait. However, if you do decide to pay for the call and dial directly you have no assurance of getting an immediate connection. The best procedure would seem to be to use an 800 number and a telephone with an automatic redial feature.

Your local telephone utility may have booklets available that will help in making long distance calls or in traveling. These are the *Personal Directory for International Dialing* and *Getting Around Overseas*. These books can be ordered by telephoning 800-874-4000. The Bell System does not publish a free directory of 800 numbers.

TELE-TAX

Tele-Tax is a telephone service offered in many areas by the Internal Revenue Service. It provides recorded information tapes on about 140 Federal tax-related topics.

You must use a pushbutton phone to call Tele-Tax. The IRS will supply you with a list of the phone numbers and the tax topics they have on tapes. Just write or call for IRS Publication 1163 (Rev 9-82). You can call Tele-Tax 24 hours a day, 7 days a week. But if you call from outside your local dialing area you will need to pay a long distance charge. Some of the tapes are also in Spanish.

HOW TO SAVE ON COUNTRY-TO-COUNTRY CALLS

Unless you have made some international telephone calls you may have the idea such calls must always be operator assisted. But quite often you can make an international call by using direct dialing and in so doing have a savings just as you have when using direct dialing domestically.

To make an international telephone call you must dial what may seem to be an unusually large series of numbers. The first number to dial is the International Access Code. This is then followed by the Country Code and, then by the City Code. Finally, dial the telephone number you are trying to reach. If you don't have the numbers required by these codes, ask your local operator. Your local telephone company may also have literature supplying information on international calls. If so, ask for a copy of this no charge booklet.

While you are speaking to your local utility operator, getting the Access, Country, and City codes, find out when the lowest international telephone call rates apply. It will probably be from 5 pm to 5 am. But before you begin to make your calls take into consideration the time differences that exist. Thus, countries in Europe are anywhere between 4 to 6 hours ahead of us in time, depending on where you are located in the U.S. and the location of the European country. European countries are east of the U.S. and so are ahead of us timewise. But countries and cities to the west of the U.S. may have a time difference of as much as 11 to 14 hours.

HOW TO DIAL MOST INTERNATIONAL CALLS

You can dial internationally in two ways — direct dialing and operator assisted. As in the case of domestic calls, direct dialing is the more economical method. However, direct dialing may not be available from your phone area, so if you have any doubt about it, check with your local telephone operator.

To make a direct-dialed international call, follow this sequence:

1. Dial the International Access code. This number is 011.
2. Dial the Country Code. This code is either a two- or three-digit number and may be listed in the white pages of your telephone directory. For countries such as the United Kingdom it is 44, for Japan it is 81, for Greece it is 30, and for Costa Rica it is 506.
3. Now dial the City Routing Code. This will be a number having 1 to 5 digits. For London (England) it is 1, for Florence (Italy) it is 55, for Hiroshima (Japan) it is 822, for Innsbruck (Austria) it is 5222. Some countries, such as Costa Rica, do not have City Routing Codes, nor do they require such a code. If your telephone

directory does not list the Country and City Routing Codes, dial o
and tell the operator the city and country you want to reach. In
some instances you may not be able to make a direct-dialed call
and will need operator assistance. This will increase the cost of the
call, but in this case you will have no choice.
4. After dialing the City Routing Code, dial the local telephone
number of the person you are calling.
5. If you are using a tone-type phone, depress the # button.

Example:

Assume you want to place a telephone call to a number in
Frankfurt, Germany and this number is 12-2456. The call would proceed in this way:

International Access Code	Country Code	Routing Code	Local Number	#
011	49	611	123456	Used only with Touch-Tone Dialing

OPERATOR ASSISTED INTERNATIONAL CALLS

Dialing an operator assisted call internationally is almost the same
as making a direct-dialed call. The only difference is that 01 is used as
the International Access Code instead of 011. After you have finished
dialing the call, the operator will come on the line and will ask questions, possibly the name of the person you are calling or your credit
card number. You will need an operator assisted call, for example, if
you want to make a person-to-person call, instead of a station-to-
station call, or if you are calling collect, or want to make use of a
credit card, or you want the long distance telephone call you are
making billed to a third party.

When making an international call it may take as long as 45 seconds for the routing of your call to be completed, although it is generally much less than this. If you do have what may seem to be a long
waiting period, do not jiggle the hook since you may disconnect the
line.

Reduced rates for night and Sunday calls apply to both direct dialing and operator assisted and person-to-person calls. For some countries reduced rates apply on Sunday only and for others reduced rates

apply nights only. For further information, telephone the Service Center of your local telephone company.

VARIATIONS IN INTERNATIONAL CALLING

You need not follow the international calling sequence for certain countries outside the U.S. This includes calls to the Bahamas, Bermuda, Canada, Mexico, Puerto Rico, and the Virgin Islands. For these countries, make your calls just as you would a long distance call in the U.S. All you need do is to dial the digit 1, followed by the Area Code and then the telephone number.

OVERSEAS TELEPHONE RATES

At one time there was a 3-minute minimum for overseas calls. You can now dial a 1-minute minimum call around the world, at any time, day or night.

Table 2-2 supplies overseas rates for countries which you can dial. Remember, however, that rates are always subject to change. If you want to verify the rates or if you need further information you can dial the Bell System International Information Service by calling 1 800 874-4000. This is a toll-free service.

USING A TELEPHONE CREDIT CARD

Telephone credit cards have had a name change and are now known as Calling Cards. They have always had the advantage of letting you make phone calls without the need for searching for coins or the worry you might not have enough of them or the correct denomination. Further, if the Calling Card is supplied by your company, billing for the calls you make is charged directly to that company, thus having the added advantage of relieving you of the responsibility of keeping a record of the charges. And a Calling Card does confer a certain amount of status in the executive pecking order. The disadvantage of these cards is they always have required operator assistance, thus putting them immediately into the most expensive telephone billing category.

In more areas around the country, you can now place Calling Card calls without help from an operator, but with pushbutton phones only. All you do is dial 0, plus the number you are calling. Where Calling Card service is available you will hear a special tone. When you hear this tone dial your 14-digit Calling Card number. You can dial just the last 4 digits when the first 10 are the same as the number you are calling. There will be no delay in reciting the number to an operator.

With the Calling Card you can make consecutive calls without re-dialing your Calling Card number. After you have completed your first call and want to place another, don't hang up the receiver. Just press the number button (#) and dial the next number you want to reach. Repeat this procedure for each additional number you want to call on your Calling Card.

Because these calls are not operator assisted, they cost less than making calls using Credit Card techniques. Further, you save time by not repeating the number to an operator.

LINE AND/OR EQUIPMENT CHARGES

The telephone lines between the Central Office of your telephone company and your home are the property of that company. They have the right to make changes in their lines or equipment. You do use the lines but you rent them only and only for the time you use them. You do not have a proprietary interest. In the event the telephone utility decides to make line or equipment changes they will notify you, in writing, of their intentions. Their purpose is to make certain you have uninterrupted telephone service.

CLEANING THE TELEPHONE

To clean a telephone all that is needed is a mild detergent and a damp cloth. Some telephones, such as those that are all metal, do not even need that for simply wiping the telephone with a cloth is enough. Do not use abrasives, scouring powder, steel wool, or solvents of any kind. These can damage the outside shell and while this will not interfere with the functioning of the telephone, it can spoil its appearance.

Keep the internal wiring of the phone away from any liquids, including water. There is no reason for cleaning the interior of the telephone. Nor is there any reason for you to tamper with the internal wiring. FCC approval is based on the construction of the telephone in the form in which you buy it but that approval does not extend to any changes you may contemplate. FCC approval is not a blanket endorsement of the phone plus changes, but simply approval of the phone as it is at the time you bought it.

CHAPTER 3

TELEPHONE TYPES

When autos first became popular, there was an ancient joke to the effect that they could be bought in any color, provided it was black. The same could be said about telephones since for many years they were available only in a limited number of colors and styles. There is no question that our telephone service was — and still is — remarkable in its performance and achievement, but now that the manufacture of telephones is no longer a monopoly practice these devices have blossomed into an astonishing variety.

TELEPHONE STYLING

You can now have telephones in almost any color, including black, tan, brown, white, blue, green, yellow, red, and these are just a few. Phones also come in just about every material: plastic, metal, leather, wood, and material combinations.

Telephones can also be had in a variety of shapes, sizes, and designs. They can be antique, modern and avant-garde, open, or concealed. And there are dedicated telephone stores whose only products are telephones, that can let you have just about any telephone you want and which offer and/or display a large selection.

THE TELEPHONE SYSTEM

The language of telephones is changing. At one time a reference to a telephone meant a unit, either single or two piece, with a dialing pad containing a combination of numbers and letters, or a rotary dial, also with numbers and letters. But with the increase in telephone sophistication, a telephone can now mean either a basic unit or a telephone having a small or large number of features.

COMPONENT VS INTEGRATED PHONES

A component telephone system consists of a basic telephone to which one or more accessories can be added, accessories whose purpose is to increase the number of available features. An integrated telephone is one that has all the desired features as a part of the telephone itself.

The advantage of a component telephone setup is you no longer need to replace your present telephone simply because it is lacking one or two features you consider desirable. You can update it (but not always) with an accessory, depending on what features you want.

What we have, then, is no longer a telephone, but a *telephone system*. A telephone, whether consisting of a basic unit or a basic unit plus accessories, is better described as a telephone system.

BASIC TELEPHONE TYPES

There are four basic types of telephones, considering the phone only from its outward appearance. These include Basic and Premium phones (Fig. 3-1), decorator phones, Advanced and Modern (Fig. 3-2). Basic and Premium are probably the most familiar types for these are the phones that have been rented from telephone companies. Decorator phones include nostalgic phones such as the European cradle phone and the early American candlestick. Some of the decorator telephones are extremely elaborate, are often finished in gold or brass, and may be supplied with a marble base, stand or pedestal. They are also probably among the most expensive types. Advanced/Modern phones are generally characterized by having a number of unusual features and are often made in a one-piece style.

ANSWER ONLY TELEPHONE

If you have one or more extension telephones (described earlier in Chapter 1) in your home, but do not use them for making outgoing

Fig. 3-1. Basic telephone (upper drawing) and premium model.
(Courtesy GTE, Consumer Sales Div.)

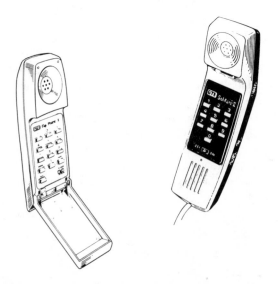

Fig. 3-2. Advanced and modern telephones are often one-piece styles.
(Courtesy GTE, Consumer Sales Div.)

calls, you are making unnecessary payments to your local telephone utility.

A way of avoiding this charge is to use an answer only extension telephone. A unit of this kind does not contain a dialer or ringer. You can plug it into any modular or standard receptacle. You are warned of an incoming call by a flashing light on the telephone. When this happens, simply move a switch on the telephone to its "on" position. After the telephone conversation is completed, turn the switch to its "off" position and replace the telephone in its cradle.

One-piece telephones, such as the unit shown in Fig. 3-3, are sometimes referred to as extension telephones. This means they are used as supplementary telephones and often have as many features (and sometimes more) as the main phone. Thus, the telephone shown in Fig. 3-3 has pushbutton dialing, a redial button, a chime ringer, on/off switch, mute, call light, and a hang-up cradle. It also has call storage which enables 24 frequently called numbers to be stored electronically and retrieved for dialing at the push of a button.

An extension phone designed for answering purposes only generally has far fewer features.

Fig. 3-3. Fully equipped extension phone. *(Courtesy Phonesitter)*

ROTARY DIALING VS PUSHBUTTON

Telephone lines are designed for two types of phones — the slower rotary type and the faster pushbutton unit. Telephone utilities make money in two ways from pushbutton phones. Because the pushbutton phone is faster it has quicker access to telephone lines. And utilities charge more for the rental of their pushbutton phones. Telephone companies refer to their pushbutton phones as Touch-Tone®. These two words are proprietary and belong to AT&T. Comparable phones are available under a number of other names from competing manufacturers.

ONE-PIECE TELEPHONES

Telephones can be either one-piece or two-piece types. A one-piece telephone is a unit that is completely self-contained. It is equipped with a telephone pad, plus a receiving and transmitting unit. The one-piece type can be hung from a wall or nested in a cradle. The cradle contains no equipment of any kind but is simply used for nesting the phone which is automatically disconnected when placed in its cradle. The photo in Fig. 3-4 shows a modular one-piece telephone with its base or cradle. It is referred to as modular to indicate the type of connector it uses for attachment to the telephone lines.

A one-piece telephone is so called since it contains its microphone and speaker in the same housing. It can be suspended from a hook or mounted in a holster or a cradle. The action of mounting the unit actuates the telephone's on/off ringer switch. As soon as the phone is removed from its cradle or holster, the switch is automatically released and a dial tone can be heard. The telephone can also be put face down on any flat, hard surface. This will also turn the ringer switch off.

The ringer switch (Fig. 3-5) can be set manually to its off position. This will disconnect the telephone from the line and no calls will be received, a much better action than removing the modular plug from the telephone jack.

The telephone cord length supplied with such units is sometimes a partially coiled cord that can stretch up to about 14 feet. When buying a phone of this kind make sure it can be used with any one of the Discount Long Distance Services.

The rightmost pushbutton at the bottom of the telephone pad,

Fig. 3-4. Modular one-piece telephone with its cradle. Some one-piece phones are suspended from a hook; others are simply placed on any flat surface.
(Courtesy Phonesitter)

marked with the symbol #, is a memory redial button. It automatically stores the last number called and at the touch of this button automatically redials your call.

Phones of this kind can be obtained in a number of different colors, including beige, brown, white, red, gold, and silver.

TWO-PIECE TELEPHONES

There are two basic types of two-piece telephones. The one shown earlier in Fig. 1-5 in Chapter 1 is a candlestick type with nostalgic appeal. To maintain the fiction of telephones of a bygone era, the pad of the telephone in Fig. 1-5 resembles a rotary dial but is actually a pushbutton type.

The hookswitch is mounted in the vertical stem of the telephone and is used for connecting or disconnecting the phone to the tele-

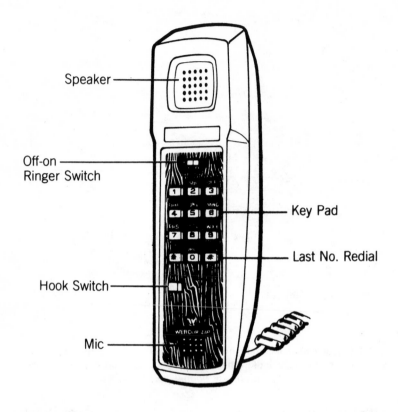

Fig. 3-5. One-piece electronic telephone. *(Courtesy Webcor Electronics Inc.)*

phone line. This switch operates automatically, closing the circuit to the phone lines when the earpiece is lifted, opening that circuit when the phone is replaced.

The earpiece contains the receiver, while the transmitter is at the top of the telephone stem. One problem with telephones of this kind is the possibility of acoustic feedback when the earpiece is brought too close to the mouthpiece.

Fig. 3-6 shows another type of two-piece telephone. With this telephone the dialing pad is mounted in a base, and so is the hookswitch. The earpiece and the mouthpiece (receiver and transmitter) are housed in the same shell. Lifting this housing closes the hookswitch automatically, permitting the dial tone to be heard.

Fig. 3-6. One type of two-piece telephone *(Courtesy Panasonic Consumer Home Electronics Group)*

TELEPHONE LIGHT SYSTEM

The usual method of telephone ringing is through the use of a bell (or bells), gong, or a set of chimes. Sometimes these sound-producing devices aren't wanted, since, under certain circumstances they can be disturbing.

Some telephones come equipped with a switch so the ringing element can be turned off, or you can get an accessory device that will do the same thing. Some phones are equipped with a lighting system in addition to a ringer bell. A red light shows when someone is calling. This feature lets you take a call in a bedroom, for example, without disturbing any other occupant. A green light glows when the phone is in use.

HANDS-OFF TELEPHONES

As its name implies, a hands-off telephone is one that does not require lifting of a handset, as in the case of two-piece telephones, or the entire phone, as in the case of one-piece units.

Hands-off telephones are equipped with a built-in speaker and the unit permits both dialing and conversation to be done without picking up the phone. Hands-off telephones can range from simple, unsophisticated units whose only feature is the hands-off characteristic. Others, more diversified, have a last-number redial memory that can accommodate a number up to 21 digits, an especially convenient feature when the user subscribes to a long distance service that utilizes long numbers.

Some hands-off telephones have other features that can include a selectable 4-level volume control with LED (light emitting diode) indicators, a quick touch call answer (which means the phone can be answered by touching any numerical button) and a mute button that allows conversation with a person in the same room without being overheard on the phone.

Some phones in this category are ac line powered, others use batteries. Fig. 3-7 illustrates a speaker phone. Like other telephones hands-off types can have a varying degree of sophistication. This unit has a 10-digit fluorescent display that shows what number has been dialed, whether that number has been handled manually or achieved with the built-in automatic dialer.

Fig. 3-7. Integrated Speakerphone dialer. *(Courtesy American Telecommunications Corp.)*

Some include a 41-number memory and each number can be up to 22 digits, useful for long distance and international calling. It may have a programmable pause to detect tones, so that the phone will wait to hear a second dial tone, or a computer tone, when utilizing a long distance calling service such as Sprint or MCI. It can also have a programmable lock to prevent unauthorized use of the phone and its programming.

THE DOODLE PHONE

Some telephones come equipped with a pad of paper, plus a pencil or pen held in a clip at one side of the phone. This eliminates the need for rushing to find paper and pen to take down messages.

ONE-PIECE AUTOMATIC DIALER

The telephone illustrated in Fig. 3-8 has the capability of storing and automatically dialing up to three important telephone numbers, each at the touch of a single button. These numbers could be for emergency use, or could be any three of those most frequently called. This telephone also features a save-any-number-dialed memory which lets the user store a phone number which was busy or un-answered and automatically redials it at the push of a single button.

While a phone of this kind is practical for any user, it is of particular importance for those who are disabled, or are ill, or who for some reason must be able to reach an emergency number without the need for full dialing. It is also useful as a security telephone since depressing a single, selected digit can make a quick call for help.

THE FLIP PHONE

Another recent type of telephone is the Flip Phone®. The entire unit is just one small, compact piece that flips open when picked up, from which it derives its name. Like many of the newer telephones, this unit has an automatic redial feature for it will dial the last number you called simply by pressing a single button marked "re-dial."

Automatic redialing is extremely helpful if you keep getting a busy number. This feature lets you "sit on top" of the busy number until

Fig. 3-8. Telephone equipped with a memory for storing up to three important numbers. *(Courtesy Code-A-Phone, Ford Industries)*

the telephone line has cleared, just by pushing the one button instead of constantly redialing.

The telephone also has a mute button. This lets you talk to someone who is in the room with you without the person on the other end of the line being able to hear your conversation. You could, of course, cover the mouthpiece with your hand but the mute pushbutton is more convenient.

Still one other feature is the ringer control. You can switch it off, or set it to either a low or high position. With the ringer set to "off," you can still make outside calls, but you can have an extension phone if you do not want to be disturbed at any time.

Note the difference between a ringless phone and an extension phone. You pay for the privilege of using a ringless, but not necessarily for an extension. A ringless gives you a continuing option for you can have it ring or not as you wish. And you can also use it for making outside calls. An extension phone, depending on the type, may not incur a telephone charge.

THE INTERCOM

An intercom (abbreviated from intercommunications) is an in-house or in-office communicating system. It is not connected to telephone lines and its use does not appear on any telephone bills.

Intercoms offer a number of advantages. They are fairly economical, do not use telephone lines, and, in a sense, represent a miniature telephone system. With the help of an intercom you can communicate from one room to another room, from one room to several rooms. Like telephones they have varying degrees of sophistication. They can be used person to person or you can use the intercom to have a conference. They use very little power and so are energy economical. They do not require dialing, and communications, assuming both parties are available, is immediate.

Intercoms operate simplex. This means that when one party talks the other listens. Transmission and reception is handled by a push-to-talk button. For those accustomed to the duplex functioning of telephones it is an awkward method. But it does simplify the design and construction of the intercom since the same component, a dynamic speaker, can be used both as a microphone and as a speaker. And, since the operation is simplex there is no chance for unwanted audio feedback, a condition that results in howling.

Intercoms can be battery operated and such units can be carried from room to room. The battery used is generally a 9-volt transistor type. Intercoms can also be powered by the ac line and while this does eliminate the portability aspect it does reduce the need to replace batteries. But even with the ac operated types, the units are so light in weight they can easily be carried from room to room. The problem then becomes one of finding a convenient outlet.

An intercom can be an independent unit, but it can be combined with and used as part of a telephone. It can be an integral part of a telephone setup, or it can be available as a separate unit with provision for connection into the telephone system.

Fig. 3-9 illustrates a combined telephone/intercom. As in the case of stand-alone telephones, the telephones used in conjunction with intercoms can have many desirable features. This one has touchbutton dialng and manual or voice (hands-free) operation. Other features include call transfer, paging, and system busy signal.

Fig. 3-10 shows a two-channel fm wireless intercom that is not associated with a telephone. It works simply by connection into the ac line. Since it is simplex it uses a touch call plate and a touch talk

Fig. 3-9. Intercom associated with a hands-free telephone.
(Courtesy Fanon-Courier Corp.)

plate. This is a convenient feature since it does not require pushbuttons. Additional units can be added to increase the number of stations.

It is possible to use the ac power lines in a home or business for carrying on telephone conversations. No special telephone wiring is necessary. All that is required is to plug the units into the nearest ac outlet. As long as both telephones are used within the ac lines supplied by one power transformer it is possible to have telephone usage. It is safe to assume a home or an office is serviced by one ac power transformer. This is not likely to be the case for widely separated homes or businesses.

With this arrangement, one unit can be used as a master but it is also possible to have duplex operation between one master and a number of remote stations, as indicated in Fig. 3-11.

The power lines used can be 110/117/220 or 240 volts, 50 or 60Hz. The units contain a miniature fm transmitter with a circuit which modulates the voice and uses the ac line as a convenient wired carrier. The base station uses a tremolo tone to reach each of the remote stations.

When calling from the master to each remote station it is necessary only to push one of the selector switches on the master unit. The called remote station will ring, just as in the case of a regular tele-

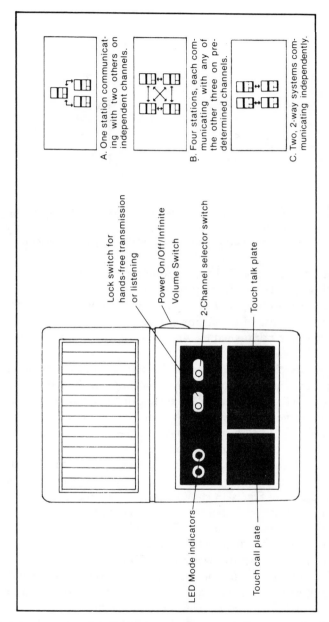

A. One station communicating with two others on independent channels.

B. Four stations, each communicating with any of the other three on predetermined channels.

C. Two, 2-way systems communicating independently.

Lock switch for hands-free transmission or listening

Power On/Off/Infinite Volume Switch

2-Channel selector switch

Touch talk plate

LED Mode indicators

Touch call plate

Fig. 3-10. Intercom uses fm transmission via ac power lines. Drawings A, B, and C indicate some possible applications. *(Courtesy All Channel Products, Div. of Electro Audio Dynamics, Inc.)*

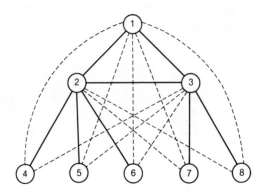

Fig. 3-11. Arrangement using three Master stations (1,2, and 3). All others are Remote stations. Solid lines show that calls can be initiated from either side. Dashed lines mean the other party can be called only from Master to Remote.

phone. It is also possible to call from a remote station to the master and this is done simply by picking up the remote handset.

There are no telephone charges since telephone lines are not used. You do make use of the ac power lines inside the home or office, but these lines are not owned by your local electric utility and so there is no charge here either.

With this setup, when used in the home, a master can be installed in the kitchen for duplex communications with a nursery, living room, or bedroom. In the office there can be duplex communications between some company executive and the factory, engineering staff, or a secretary. In a hospital a physician can use it to reach a nurse, an X-ray room, or a pharmacist. See Fig. 3-12.

Although the system just described makes use of three remote stations, it is possible to have a setup that will cover as many as eight. The diagram in Fig. 3-13 shows how four master stations are set up to communicate with one or more remote stations. As mentioned earlier, this system uses frequency modulation (fm) and the operating power is less than three watts. Signal strength is in the order of 50 milliwatts. With the arrangement shown in the drawing it is possible to select any remote station from the master station for calling, talking, and paging. The units are equipped with a press button for talking with the same button released for listening.

Note, in the drawing in Fig. 3-14 there is one master station and eight remote stations. While the master can communicate directly with any one of the remotes, any remote must work through a master unit to call and talk to any of the remotes.

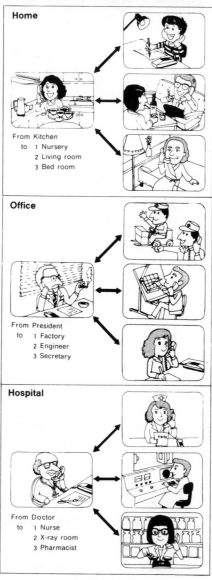

Fig. 3-12. Multistation arrangement using ac power lines.
(Courtesy Samhill Enterprises, Inc.)

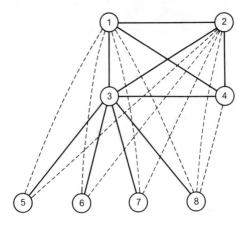

Fig. 3-13. In this setup, items 1 through 4 are Master stations; all the others are Remote stations. Solid lines indicate that calls can be initiated from either side. Dashed lines show that the other party can be called only from Master to Remote.

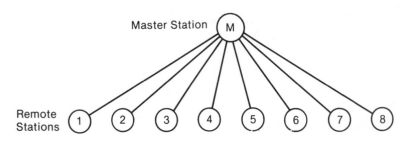

Fig. 3-14. System using 1 Master and 8 Remote stations. The Master can communicate directly with any of the Remotes. The Remotes can also communicate directly with the Master but must go through the Master to be able to reach another Remote.

Various other arrangements are possible, but the greater the number of master units the greater the flexibility. As shown earlier in Fig. 3-13 there are four masters, numbered 1 through 4. Those numbered 5 through 8 are remote units. The solid lines show the other party can be called either from the master or from the remote. In other words, the calling action can be initiated from either side. The dashed lines indicate the other party can be called only via a master.

BUDGET PHONES

Some elaborate telephones are fairly expensive and can cost several hundred dollars. But you can also get a telephone at rather low cost. Although a budget telephone can cost less than $25, it can have a number of desirable features. It could have a redial button identified by the letter R and clearly marked redial. This is the pushbutton at the extreme right on the bottom row. Depressing this single button lets the telephone redial the last number you called, and it will do so automatically. The unit might also have an on/off ringer switch. This cuts off the ringer but leaves the telephone operational. The telephone should also be compatible with all telephone systems, tone or rotary.

REBUILT TELEPHONES

There is a growing market in used and rebuilt telephones and these are sometimes offered at very low prices. However, the emphasis will be on the telephone's numerous features, not on the fact that it is a used machine, and so this downplayed aspect can be overlooked. Sometimes these units are sold without any warranty, or if there is, a warranty would only be applicable to the original owner. Some are sold on a "trial" basis, generally for a period of about 10 days. On some of these trial offers, you will not get your money back when you return the telephone, but rather another telephone to take the place of the one you originally bought. Vendors of such telephones take advantage of the fact that running back and forth with a telephone is awkward and inconvenient. And they also know that operating on a "swap-only, no money return" basis means you become a captive customer.

If you buy by mail you will need to read the advertising solicitation carefully. Phrases such as "good as new," "completely reconditioned," or "would sell for twice this price if new" all mean the same thing — the telephone is a used unit. Whether the telephone has indeed been reconditioned, or made as good as new is anybody's guess, including yours. There is no way of verifying the claim.

Not all budget priced telephones are so-called rebuilts. The brand-new unit illustrated in Fig. 3-15 retails for less than $20 and has a number of desirable features. These include an automatic on/off hookswitch. A red LED light on the back of the phone shows

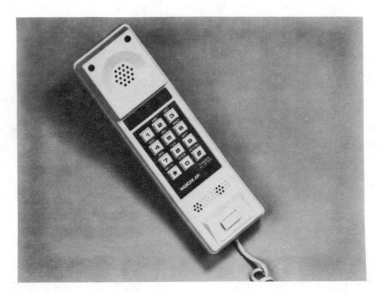

Fig. 3-15. Budget-priced telephone. *(Courtesy Webcor Electronics, Inc.)*

whether the telephone is on or off. The unit is also equipped with antiskid pads to keep the phone from slipping off a desk or table.

The telephone is supplied with a three-position ringer control switch to let you turn the ringer off or to adjust the volume to a low or high position. A microphone-mute control prevents the person on the other end of the telephone line from hearing conversations or noises occurring in the room you are calling from. Like a number of other, but more expensive, phones this unit has a last number redial key which dials the last number called if the line was busy and you want to call it again. The unit can be put on a table or wall, or mounted with an optional holster.

TELEPHONE PRICES

No manufacturer of telephones or related equipment, such as accessories, has the right to establish retail prices. The best any manufacturer can do is to supply suggested retail. This means a vendor of telephones and telephone equipment sets his own pricing schedule and this depends on competition, the desire of the dealer to make a sale, his need to move equipment to make room for new models, or to improve his cash position.

For this reason you may find identical models of telephones being sold at various prices, so for this reason it is advisable to do some shopping. As a general rule, the greater the number of features the more expensive a unit will be. Cost is also determined by the materials of which a phone is made and its style.

It is also advisable to get the best warranty terms possible and to learn, at the time of purchase, where telephones can be shipped for repairs. Repairs are generally handled by the manufacturer or he may set up one or more repair depots. Some individuals are also establishing phone repair shops, just as we have radio and tv repair stores.

If the telephone you bought is out of warranty you will not only be expected to pay for parts and labor, but for shipping costs in both directions.

ONE-PIECE/TWO-PIECE TELEPHONES

One-piece and two-piece telephones were described earlier, near the beginning of this chapter. Usually these are two distinct types but it is possible to have a telephone that combines the features of both.

The telephone illustrated in Fig. 3-16 is a one-piece unit but it is made into a two-piece type by being supplied with a heavy cradle so the phone can be mounted on the wall; placed on a desk; or on a soft surface such as a sofa, bed, or carpet.

Quite often a one-piece telephone is supplied with a hook so it can be suspended from a wall, or it may simply be put on a desk, but the telephone in Fig. 3-16 enables the user to put the phone in the most convenient location. This can be virtually any surface. The cradle is equipped with anti-skid pads to prevent it from slipping.

This telephone has outpulse dialing that works with either rotary or tone dialing systems. It has an automatic on/off switch and a three-position ringer control switch that can be set to off, low and high volume. There is a last number redial key which, at the push of a button, dials the last number back if the number is busy or if you want to call the person again.

MUSIC ON HOLD

Some telephone systems used in offices and homes play synthesized music when a caller is put on hold. Music or melody on hold has

Fig. 3-16. One-piece/two-piece telephone. *(Courtesy Webcor Electronics, Inc.)*

several purposes. It informs the caller the phone line is still connected. It also entertains the caller until the call is put through. Music on hold is automatically terminated when the called person answers the phone.

The telephone shown in Fig. 3-17, is a one-piece unit, and is designed for people who have two telephone lines in their homes, allowing three-way conversations and conference calling. It allows the user the convenience of using two telephone lines without having to walk to another location. This telephone automatically hangs up when you put it down. The ringer can be turned off and it also has a last number redial capability.

ON-HOOK DIALING

With the usual telephone it is customary to lift the phone from its cradle prior to dialing. With the phone shown in Fig. 3-18, however, it is possible to dial through the built-in speaker with the telephone on-hook. This feature can be defeated and the phone can be operated in the usual manner.

This telephone has switchable dialing, from rotary to touchtone, for use with MCI and Sprint. It also has programmable memory lo-

Fig. 3-17. Telephone supplies MELODY ON HOLD™. *(Courtesy U.S. Tron)*

Fig. 3-18. This unit can be dialed with the telephone resting in its cradle.
(Courtesy Webcor Electronics, Inc.)

cations able to store up to 16 different numbers each with a 16-digit capacity.

PRESSURE SENSITIVE PADS

Most telephones are either rotary dial or pushbutton types. There are several variations of pushbuttons: one is the regular, protruding kind, while another, as shown in Fig. 3-19, uses pushbuttons which are practically flush with the surface of the telephone. Still another is the pressure sensitive type. These do not require a push, but simply the touch of a finger.

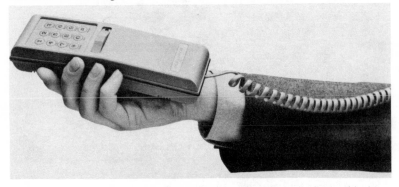

Fig. 3-19. This model has pushbuttons that are practically flush with the surface. *(Courtesy SPS Industries)*

DESKTOP PHONE

The two-piece telephone shown in Fig. 3-20 is a style that has been available for a long time. But even these phones are undergoing a change for they are now equipped with more features. Thus, they are available for either true tone dialing or outpulse dialing for use with either rotary or tone dialing systems. You can get a desktop phone with a mute button which enables you to hear the person on the other end of the line without his (or her) hearing the conversation or noise in the room where you are. It also has a last-number redial key.

WALL HOLSTER

There is always one question that arises with a telephone and that is where to put it. If the phone has a base with a cradle then any flat

Fig. 3-20. Desktop telephone *(Courtesy Webcor Electronics, Inc.)*

surface is generally suitable. Some one-piece telephones can be suspended from a hook. Some telephones, though, such as the one shown in Fig. 3-21, are supplied with a holster. This particular holster, made of clear plastic, has place for the storage of extra cord in case it is necessary to shorten the cord length. There is also room in the holster for a hidden directory of often-used telephone numbers. The telephone associated with this holster has a built-in 11-digit automatic dialer which can dial both rotary and tone dialing. It also supplies last number redial. No ac adapter is needed, nor are any batteries required.

The telephone can be programmed to store numbers up to 16 digits long. However, if you require storage of numbers longer than 16 digits, you can divide the numbers into groups and store these groups under the additional storage keys of your choice. Depending on the telephone model you select, you can store 11 or 22 numbers. Numbers are stored in the phone's internal memory and retrieved by the push of two buttons.

The telephone is equipped with a mute function, melody on hold, and a tone ringer with an on/off switch.

LINE-DISCONNECT TELEPHONE

While the telephone is an excellent means of communications there may be times when you do not wish to be disturbed. There are var-

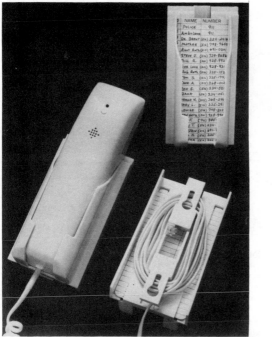

Fig. 3-21. Telephone is supplied with a wall holster. *(Courtesy U.S. Tron)*

ious ways of handling this problem. You can remove the receiver from its hook, remove the telephone plug from its wall jack, use a telephone cord that is equipped with a disconnect switch, or get a telephone that has a disconnect feature.

Some of these methods are undesirable. Removing the receiver from its hook will give all your callers a busy signal. Further, it may activate equipment in the Central Office that will produce a warning tone from the receiver. Removing the telephone plug is not recommended since the plug is not designed for and is not meant for this type of usage. Using a telephone cord equipped with a disconnect switch is preferable, but you may find it inconvenient if the cord is draped behind a desk or other furniture.

One of the features of the telephone shown in Fig. 3-22 is that it is switch equipped for turning off the ringer. It also has a muting control, a redial memory for up to 20 digits, and a music-on-hold function.

Fig. 3-22. A feature of this telephone is a line-disconnect switch.
(Courtesy Contec Electronics, Inc.)

TELEPHONE CALL SCREENING

The telephone in Fig. 3-23 lets you decide whether your phone should be on or off. A special programmable feature lets you turn your phone on for 1 to 10 hours and then automatically turns it off again. Or, it can totally silence your phone from 1 to 10 hours.

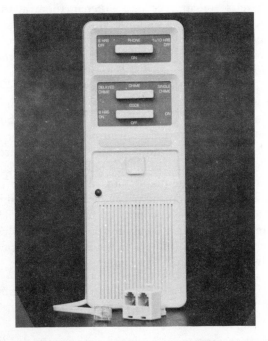

Fig. 3-23. When this phone rings, one of four special chimes announces who is calling. *(Courtesy Chrono-Art, Inc.)*

When more privacy is wanted, wrong numbers and crank calls are screened by codes that ensure selected friends and family members can get through. The phone in Fig. 3-23 has four codes. Each of these has its own distinctive chime, letting you know who is calling or who the call is for.

There are a number of manufacturers who make these phones, with each having a varying number of codes (Table 3-1). The maximum number of codes presently available is 12. With some of these phones you can change the codes, while with others the code is fixed. And, depending on the phone selected, you can have a choice of chimes and a light, an electronic warble, or the usual type of ring. One of these types has four different rings plus a display of a digital code.

The telephone shown in Fig. 3-23 also has an unusual feature. Nobody wants the telephone to blast, especially if you are standing right next to it. With this telephone the first three rings are at normal volume, with the sound level selected by you. Starting with the 4th ring and successive rings, the phone will work at full volume. The advantage here is you may be in a different room at the time the phone rings, or the ambient sound level may be too high to let you hear the phone. The fourth and following rings should be loud enough to override these sounds. You can return to normal volume on all rings simply by turning this feature off.

Still another feature of this telephone is hold, mentioned earlier in connection with another unit. Thus, if you have two or more telephones in your home you can answer any one of them but you may want to continue your conversation on one of the other phones. In that case all you will need to do is to press the hook twice and hang up. This action will put your phone on hold. You can then walk into another room, pick up the phone and answer the call. This action releases the hold control on the telephone you answered originally.

You can also use the hold control to cut down on the time you spend on an unwanted call. Thus, if you have a persistent person calling via phone, put him (or her) on hold. A timer releases hold and automatically hangs up after four minutes.

MEMORY TELEPHONES

Many telephones are now equipped with a memory (mentioned briefly earlier in connection with Fig. 3-7), but not all memories are

Table 3-1. Comparison of Features of Telephone Call Screening Products

	PHONE CENSOR	PHONE CENSOR PLUS HOLD	FOX FONE INTERCEPTOR	THE BUTLER	PRIVE-CODE
Number of Codes	2	4	1	1	12
Type of Code	Ring Pattern from Touch Tone or Rotary phone	Ring Pattern from Touch Tone or Rotary phone	Digital from Touch Tone only	Digital from Touch Tone only	Digital or Voice Interrupt from Touch Tone or Rotary Phone
Can you choose which codes are active?	No	Yes	Only one code	Only one code	Yes
Can Codes be changed?	No	No	Yes	Yes	Yes
Is the phone answered and the caller billed if you're not home?	No, caller is not billed.	No, caller is not billed.	Yes, caller is billed.	Yes, caller is billed.	Yes, caller is billed.
How are you informed of a call?	Choice of 2 Chimes & a Light	Chimes & a Light	Electronic Warble	Electronic Warble	Normal Ring
Are you informed of who's calling or who the call is for?	No	Yes, 4 different chimes—one for each code.	No	No	Yes, 4 different rings & display of digital code.
Other ring control features	Single Chime Delayed Chime Volume Adjustment	Single Chime Delayed Chime Volume Adjustment 4th Chime Louder	Volume Adjustment		
Timed features	8 hr Code 1, 2, & 8 hr off 1, 2, & 8 hr on	8 hr Code 1 to 10 hrs off 1 to 10 hrs on	No	No	No
Other Features		Hold feature for all phones on one line		Telephone included	Stores log of who called

alike in the sense they do not all have the same memory capacity. The simplest memory is last-number redial but this is a memory for one phone number only.

The telephone shown in Fig. 3-24 has a 10-number memory. For your frequently called numbers this telephone will store up to 10 of them and will automatically dial them on your command. Each of these telephone numbers can be up to 16 digits.

Most one-piece telephones have their hooker switch positioned in the front. In this location it is easy to disconnect the telephone accidentally. For this reason this telephone has its on/off switch mounted on the side, so you do not disconnect whenever you lay the phone down. Another feature is a 10-number index card inside the base for listing the names and numbers stored in the automatic dialer memory.

Office telephones generally have a hold feature, something not found in home-type phones. This unit has hold. A soft beep sounds every few seconds to remind you the telephone call is waiting. The phone automatically disconnects after being on hold for five minutes and returns your telephone line to service so additional calls can come through.

There are some telephones that have a larger memory capability.

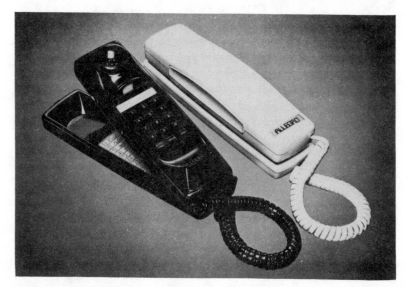

Fig. 3-24. This phone has 10-number automatic dialing and last-number-dialed memory. *(Courtesy Comdial Corp.)*

One business oriented model can store as many as 176 telephone numbers, while another, intended for in-home use, stores up to 93 local numbers and 55 alternate long distance numbers. Having a large capacity memory has distinct advantages. It eliminates dialing wrong numbers or dialing long alternate phone system (MCI, Sprint, ITT, etc.) digits.

Storing a number is as easy as remembering a name. When programming your personal "phone book" you can store names from 2 to 6 digits (like Mom or Bob) or by code numbers, just like a conventional dialer. To dial the number, just dial the name and the telephone does the rest.

Automatic Redialing of Busy Signals

Many telephones now have a redialing feature, but not all redialing methods are the same. A phone is available that hears and interprets busy signals, rings, dial tones, and computer tones. If a number is busy, for example, the phone will automatically detect that busy signal and will redial the number 10 times the first minute, 5 times during the next 10 minutes at two-minute intervals and ten every 10 minutes for up to two hours. The unit dials at the fastest speed allowed by the FCC.

Automatic Redialing of Unanswered Phones

The same unit described above will also redial an unanswered phone every 10 minutes for up to 10 hours and will send you a signal as soon as someone picks up the phone.

Three-Number Memory

The amount of memory you need in a telephone depends on your memory requirements. While a telephone having a 40-number memory is impressive, this feature is wasted if you just require a memory for a few numbers only. Usually, the larger the memory the higher the selling price of the telephone.

If your telephone memory requirements are modest, you can get a telephone such as the one shown in Fig. 3-25. The press of a button dials any of three phone numbers that have been programmed. Pressing the redial button automatically redials the last number called.

The telephone is compatible with all telephone systems (tone and rotary). You can "hang up" the telephone just by putting it on any

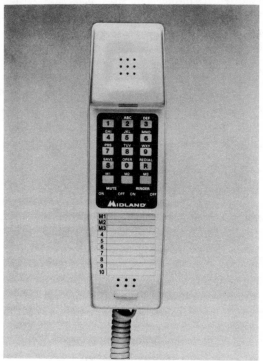

Fig. 3-25. Telephone equipped with a 3-number memory.
(Courtesy Midland International Corp.)

flat surface. Since it is a one-piece phone it has no hookswitch, so the switch mounted on the back of the telephone is used for disconnect purposes.

An on/off ringer switch cuts off the ringer but leaves the telephone operational. There is also a mute switch for private conversations. As a further convenience, a color-coded index on the front panel lets you record the three programmed phone numbers plus seven other important or emergency numbers.

Ten-Number Memory

The one-piece telephone illustrated in Fig. 3-26 has a 10-number capacity. The fact that a phone is memory equipped does not mean there is any digit limitation on regular dialing. Since this unit is a one-piece component you can hold it in one hand while dialing. It features automatic last number redial, an in-use LED indicator, an

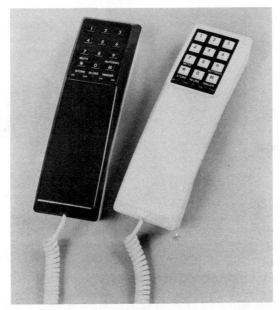

Fig. 3-26. Telephone equipped with a 10-number memory.
(Courtesy Mura Corp.)

on/off ringer switch, and a mute button for private conversations. The dialing is universally compatible, hence can be used with either rotary or tone systems.

Large Memory Telephone

While many telephones now have a last number dialed memory, some such as the unit illustrated in Fig. 3-27 may have as many as 41. With this unit you can make calls with just the touch of a single button, with the telephone dialing those numbers for you automatically. There are two directories built in, making it easy to keep personal and business numbers separate. A switch reveals the ones you want.

Recording new numbers, or changing numbers, is just a matter of a few seconds. Just press the record button in the lower left corner of the unit, press the appropriate call button and dial the number. You can program up to 22 digits, even international long distance.

You can also place a call without picking up the phone. When your call comes through just lift the receiver to talk and the speaker turns off automatically. To use the monitor speaker, just depress the speaker button.

Fig. 3-27. Voice Express 41 integrated speakerphone dialer has 41-number memory. *(Courtesy Comdial Corp.)*

A visual LED display gives an automatic readout of the number dialed. It can also recall any number stored in the telephone's memory. With the digital call duration timer you can keep track of calling time. The memory is protected by 4 alkaline AA batteries and should have a life of about one year. An L appears in the LED display to let you know when batteries are low and need replacing.

TELEPHONE STYLES

Telephones are available in just about every conceivable style, shape and design. Often, though, you will need to make a decision as to whether you want certain features or prefer a decor-type phone you find appealing. Decor phones are not usually characterized by having numerous features.

Two-Piece Decor Phone

The telephone shown in Fig. 3-28 is a highly decorative unit using what appears to be an older style rotary dial. However, the dial does not rotate and, instead of having finger holes, is equipped with

Fig. 3-28. Two-piece decorator type telephone.
(Courtesy Comdial Corp.)

pushbuttons. Unlike telephone pads that are ordinarily rectangular, this pad is circular. Phones of this kind are also available with a rotary dial. The base on which the telephone rests can be made of any material — plastic, wood, leather, onyx, metal, glass, etc.

Miniature Phones

The size of a telephone is not standardized and so miniature telephones, such as the one shown in Fig. 3-29, are available. The unit has a touch-sensitive pad. With this arrangement there is no need to depress pushbuttons. Instead, just a light touch on each digit is all that is required. The unit produces tones in the same way as any regular pushbutton phone.

Telephones for Children

The telephone shown in Fig. 3-30 is a fully functional unit colored in bright green and available as a pushbutton tone type or rotary dial. The head of the frog can be moved around to create a number of poses.

Fig. 3-29. Compact telephone. *(Courtesy SPS Industries, Inc.)*

Other telephones for children are the Mickey Mouse, (Fig. 3-31), Pac-Man, Winnie-the-Pooh, Snoopy and Woodstock. These phones are also available with lamps which are 27½ inches high and come with a three-way switch and an off-white pleated shade.

Genie Phone

This lightweight phone is compact and is available in a number of different colors and materials, including leather. Some of the phones are in a single color; others have two. All have an electronic ringer with a two-level volume control and a round pushbutton tone pad. Fig. 3-32 shows the unusual shape of this line of phones.

MULTIFUNCTION TELEPHONES

The concept of a telephone as simply a duplex device permitting two-way conversation, plus some rotary or pushbutton dialing system, is gradually changing. Telephones are now expected to do much more. Thus, the multifunction phone shown in Fig. 3-33 combines the benefits of a remote activated answering machine, automatic dialer and single line telephone. (Answering Machines are described in greater detail in Chapter 7.)

Fig. 3-30. Kermit the Frog telephone. Copyright © 1982 Henson Associates, Inc. *(Courtesy Comdial Corp.)*

The handset of this phone system contains all the controls needed for placing calls, including automatic dialing, conveniently placed for use without returning it to the base unit. The automatic dialer has a 10-number memory and is compatible with alternate long distance dial systems like Sprint or MCI. Its dial modes are tone and pulse (10 pulses per second.)

The unit is actually a two-piece arrangement — the telephone itself and its associated base unit. The base unit contains a speaker monitor which sounds a dial tone, ring, answering voice or busy signal so the phone can be used with the handset on the cradle. The speaker also monitors incoming messages on the answering machine. This phone system will also redial the last number called at the press of a button.

The unit also provides the service of an answering machine. It allows one outgoing message of any length up to 3 minutes and for the second to be of any length. It uses two standard cassettes, one for incoming and one for outgoing messages and automatically switches to the second outgoing message when the incoming message tape is full. Equipped with a remote control system it can be instructed to playback, reset, skip, and repeat over the phone.

Fig. 3-31. Mickey Mouse telephone. *(Courtesy Comdial Corp.)*

Fig. 3-32. Genie telephone. *(Courtesy Comdial Corp.)*

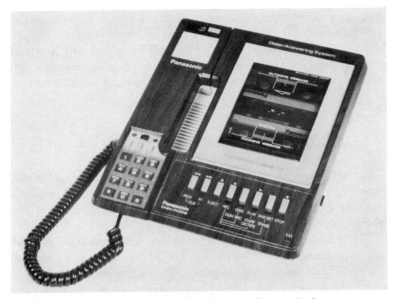

Fig. 3-33. Multifunction telephone. *(Courtesy Panasonic Consumer Electronics Group)*

Incoming messages are voice activated. The system also provides a cue and review function to simplify finding a specific message and it acts as a recorder of two-way conversations and as a dictating machine as well. An electrical locked quick-erase function rapidly prepares the incoming tape for new messages.

SUPERPHONES

The integrated telephone is one that combines several devices into the phone, such as an answering machine, automatic dialer, chime ringer, etc. But these units are directly related to telephones and integrating them with a telephone is to be expected.

However, there are now telephones that are being combined with units you wouldn't usually expect to be associated with phones. You can get a telephone that not only has a 20-number automatic dialer, but is equipped with a calculator, a digital alarm clock, and a stopwatch. Thus the telephone is making a move toward becoming an in-home center for various electronic products.

The all-in-one telephone is desirable from an economy viewpoint since it is often more economical to buy a phone that is equipped with a variety of devices rather than buying all these on an individual basis. But whether such a telephone will be convenient is another matter. You may not want a telephone in the bedroom where you usually keep an alarm clock. You may not want a calculator as part of your phone where it would be in a fixed location.

HOME MANAGEMENT PHONE

The telephone shown earlier in Fig. 3-33 is a multifunction telephone in the sense that it is a combined telephone and answering machine. But it is also possible to get a phone that is multifunctional in every sense of the word.

The unit, shown in Fig. 3-34 has six electronic systems that let you run your home from a telephone. The phone includes an intercom system with a hands-free response. The intercom section and the telephone function completely independently but can be used simultaneously. You can also use the intercom as an emergency unit to interrupt a busy telephone.

The system is compatible with user supplied systems. It can control an intrusion security alarm, a fire detection alarm, background music, and a remote door opener. With its "controller" you can turn on the lights, heat, or appliances from remote locations. You can answer the door from any telephone, screening visitors and deliveries, and open a door usng a remote door answer/open device.

The component also works as a sensitive monitor. You can babysit the sick or keep an ear on children in another room. This phone has an individual volume adjustment and is compatible with either an existing rotary or pushbutton service. If you have more than one phone in your home you can transfer an incoming call to any phone. It has automatic ring back of a call on hold.

HANG TELEPHONE

Most telephones, whether one- or two-piece types, have a switch that automatically connects the unit to the line and so no manual switching is required. The receiver may be lifted from its hang switch or the phone may be lifted from its cradle. In either instance this operates a switch that puts the phone in its operating position.

Fig. 3-34. Home-management telephone. *(Courtesy Technicom International, Inc.)*

However, it is possible to have a telephone in which the switching must be done manually. This is the requirement for a hang telephone, a unit that does not have a cradle or hookswitch. Instead, the phone is suspended by a hook which can be mounted on a wall or convenient place. The trouble with manual switching is that it is easy to forget to turn the phone either on or off.

The method of switching has nothing to do with the features of the telephone, and with a hang phone you can have as many, or as few, features as comparable units. Thus, the hang telephone can be equipped with a memory button, although some are equipped with a number of these; a last number redial button; a mute switch that lets you talk to someone in the room without the incoming caller overhearing; and an electronic ringer instead of a bell.

DUAL DIAL TELEPHONE

This telephone is equipped with a dual dialing system, with key pads placed side by side. One of these pads is for pulse dialing; the other is for tone dialing. Either dial can be used at any time. The purpose of this phone is to supply both tone and a rotary dialing capability when mixed line service cannot be avoided.

OUTPULSE DIAL TELEPHONE

The problem with rotary dial telephones is their slowness and this is inherent in the movement of the dial. To overcome this difficulty there are telephones that produce the equivalent of rotary dial pulses, but use a pushbutton key pad. But while the outward appearance has been changed so the telephone resembles a pushbutton type, its output is that of a rotary dial phone and so it is not compatible with any common carrier toll network service that normally requires tone dialing. You cannot assume every pushbutton type telephone is a tone producing type.

HOME COMMUNICATIONS CENTER

The usual in-home telephone has, for many years, been extremely limited in its capabilities. More recent phones now have a last number redial capability and some are more extensively memory equipped. But the sophistication possibilities are possibly best illustrated in the phone system shown in Fig. 3-35.

This unit has a conference speaker to allow on-hook dialing, leaving your hands free. It lets you hear the phone being answered, but the speaker automatically disconnects after you pick up the receiver. Thus, you have the option of privacy or the use of group communications, and you can go from one to the other at any time.

Another feature is programmable automatic redial which helps you get through to a busy line as soon as the other party hangs up. In doing so it will repeatedly redial the number you are trying to reach, up to 14 times at 1-minute intervals. It supplies you with 30 direct telephone lines to people you call frequently, either locally or long distance.

The rather elaborate pad of this unit is illustrated in Fig. 3-36. If you make a dialing error it is not necessary to hang up and go through the dialing process. Instead, you can backspace erase. Another button lets you stop any operation in progress, especially useful if you change your mind after dialing a number. Technically, there

Fig. 3-35. Multifunction telephone with LED readout.
(Courtesy Dictograph Corp.)

are no buttons on the pad. Instead, there is a pressure sensitive key pad using solid-state circuitry. The advantage of such a pad is there are no moving parts as in rotary pad or pushbutton units, hence less opportunity of wear.

Whenever any number you call requires an access code, you must wait long enough for that code to function. This unit permits time pauses of any length to be programmed for use in phone systems where an access code, such as 9, must be dialed to obtain an outside line.

Still another feature is an instant LED Reference Readout of numbers programmed in memory position, even without dialing. Also, with this component you can access computer services requiring more than 16 digits, and it will even dial MCI and Sprint numbers.

You can use it to transfer an incoming call to an extension telephone. You can time all outgoing calls automatically, thus cutting

Fig. 3-36. Pad of the Phone Controller. *(Courtesy Dictograph Corp.)*

down on long distance expenses. A reminder tone tells you to take note of the time once the call is completed and stores the time of the last call for later referral.

The phone is equipped with a lock to prevent unauthorized use. It provides security for stored numbers and keeps private numbers private. It has a backup battery compartment to ensure full operation during a power failure and maintains the numbers in memory storage.

MULTILINE BUSINESS TELEPHONE SYSTEM

Most home telephone systems have just one telephone line although they may have two or more phones. But with just one line the two phones cannot be used simultaneously for making outgoing calls, although both can be used at the answering end of an incoming call.

For a business, a single line is inadequate, and for a small business installing a multiline system can be a heavy expense. However, you can get a "do-it-yourself" prepackaged business telephone system anyone can install.

The system comes in four-line, six-line, and eight-line models. It can accommodate a total of thirteen six-button telephones and the phones can be added at any time. The buttons do not refer to the phone pad, but to controls adjacent to the pad and used to perform various functions such as call transfer, intercom, etc.

One of the advantages of this system is that it can be maintained and serviced by anyone without prior technical knowledge. You do not need expensive service contracts or specially trained telephone technicians.

This system can be equipped with hands-free, single-path, call announcing dial intercom. This means you can call from any telephone in the system, and talk to another person. They do not need to touch the phone, or be close to it to hear and talk to you. It also does not interfere with telephone conversations.

This system can have a hands-free announcing intercom on any phone and it can have conferencing on all lines. It can also be equipped with one "music on hold" card. This lets you connect an audio setup to your phone system and while people are on hold are listening to music, they have an assurance they have not been cut off. You could have sound supplied by an 8-track cassette deck using its endless loop cartridge to deliver your sales message, possibly interspersed with music, also while your callers are on hold.

This system is equipped with "paging ports" which permit you to attach peripheral paging equipment and control units to your phone system. Then, by accessing intercom and dialing 29 you can have an all paging function.

You can also equip the system with automatic dialing. The dialer will store up to 176 numbers and is MCI and Sprint accessible, letting you use alternate long distance services. It can continually redial busy or unanswered numbers and then ring when your party answers. And, with a single auto dialer every telephone connected to this system can use it.

An additional feature is power failure cut/through. This feature means if there is a power failure of any kind the system will revert to telephone company power (tip-ring) and can be preset to signal, dial outside calls and receive them with no interruption of service.

Telephone Master System

The concept of a telephone being used solely for voice communications is gradually changing. Instead the telephone is becoming a center, an advanced multifunction system offering automatic answering and dialing, as well as working as a duplex speaker phone, plus a control and protection center.

Fig. 3-37 shows this integrated system consisting of three components. The unit at the left is an automatic answering system equipped with dual minicassettes which can record up to 60 minutes of incoming messages. It records all incoming calls, times and dates, then automatically displays them during message playback to provide a complete and accurate call log. The automatic answering system includes a seven-channel outgoing message tape that allows two regular use announcements, plus "tape full" and "announce only." The "announce only" feature means no incoming calls will be recorded. "Tape Full" means that the tape that records incoming messages can accept no more. When that happens, a tape full message is played automatically.

The outgoing message does not need to request the time and date from the caller. Instead, it records this information automatically. When you play back your messages, the date and time are displayed along with the number of calls recorded. This sort of information is particularly helpful if you have been away for a number of days and are interested in the time sequence of your messages.

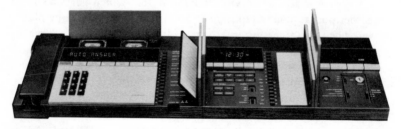

The unit is equipped with VOX or voice activation. This feature helps economize on recording tape and time. With VOX the unit only records when the caller talks. It eliminates long pauses and dial time between messages. This means more messages on a tape.

There is also a remote message indicator. If you call from the outside, the unit lets you hang up if no messages have been recorded and it lets you do so before you are charged for a call.

The other two sections of this integrated system are the Control Center and the Protection Center. The Control Center supplies instant and automatic remote control of lights and electrical appliances.

The Protection Center supplies wireless, multifunction security, including burglary alarm/deterrent, fire and smoke alarm, personal/medical emergency alert and utility failure warning. This component uses radio-frequency transmitters instead of conventional hard wiring.

TELEPHONE AIDS FOR THE DISABLED

There are telecommunications devices for the deaf (TDDs) that allow hearing impaired persons to communicate with other TDD users over the telephone network. Portable printers encode typed messages into sounds that are transmitted over the telephone network to similar units. Toll-free dialing assistance is made available through special operators.

Amplifying Handset for Public Telephones

Some public telephones now have the capability of increasing the sound volume of the voice supplied by the telephone receiver. This is done by using a slide switch or a pair of buttons on the phone. There is no need to make any adjustments to the phone after a call is completed since the sound is returned to normal after the handset is hung up.

Most of the phones equipped with an amplifying handset are those that have a high degree of usage, such as telephones in airports and bus depots. Hospitals and nursing homes are now equipped with these handsets.

There are various ways of recognizing a public telephone that is equipped with an amplifying handset. There may be a sign on or near the telephone booth. The handset may be equipped with a sound wave design on the back.

Compatibility with Hearing Aids

Most Bell System telephones are compatible with hearing aids that rely on an electromagnetic field when used with a phone. Those telephones that are not compatible can be modified and those that cannot be altered can be used with a telephone adapter. The adapter does not amplify sound, but generates a magnetic field on which the hearing-aid telephone pickup depends.

Bone Conduction Receiver

Some hard-of-hearing persons hear better by bone conduction. You can get a telephone having a receiver that transmits sound to the inner ear. In bypassing the normal hearing process, it helps those with certain types of hearing loss use the telephone.

Auxiliary Watchcase Receiver

For those who are deaf, it is possible to carry on a telephone conversation with the help of a third person. That person uses an auxiliary receiver to listen to the party at the distant end and then either repeats the words if the deaf person is a lip reader or uses sign language.

Headset Amplifier

A small plug-in type amplifier is available for those who have hearing problems. The amplifier increases the sound level being delivered by the headset receiver. The amplifier and its accompanying headset can be used with PBX consoles, switchboards, and certain telephone sets which are equipped with "jacks" for plugging in headsets. The headset amplifier is intended for switchboard operators who have impaired hearing.

Ringer Signals

Persons who have hearing problems cannot always hear the telephone's ringer, particularly if they are in a room that is away from the telephone. However, there are various devices that can be used to overcome this difficulty.

Tone Ringer

Your telephone can have a ringer that concentrates all of the sound energy in a frequency range which the majority of persons with impaired hearing can hear. Or, the phone can be equipped with a bell that is much louder than the bell normally used. Your telephone can be equipped with a bell that is much louder than the usual telephone bell. An eight-inch gong is also available. This produces a loud tone in the bass frequency range and is intended for people with severe hearing problems. It can be heard over high-level background noises and at a substantial distance.

Signalman

This device is an aid for those whose hearing problem is such that they are unable to hear the telephone ring. Any lamp, when plugged into the unit, will flash on or off each time the phone rings. When the lamp is off, it flashes on. If the lamp is on, it flashes off at each ring. An additional feature is a small neon pilot light to indicate the system is working.

For someone who is blind and deaf, a small electric fan, when plugged into the unit, will signal the phone's ringing by gently blowing toward the person.

Aids for Impaired Speech

You can get a telephone that has an amplifying handset. An adjusting wheel on the handset can be used to increase the volume of the speaker's voice. The volume can be turned down when someone with normal speech uses the handset.

Aids for Impaired Vision

If you have a rotary dial telephone you can have it equipped with a large-number dial ring. The dial ring has an adhesive back which sticks to the outside of the rotary dial. Some Bell companies also offer large stick-on numbers for telephones equipped with pushbuttons.

Aids for Impaired Motion

There are two ways in which a phone can be used by someone with impaired motion. The first is the help of a speakerphone, a telephone that does not require the lifting of a handset. If the telephone is not that type, an accessory device can be had which converts an ordinary telephone into a speakerphone.

For those who are unable to lift a handset, a headset can be used instead. The headset is equipped with an adapter cord with one end fitted to plug into the telephone. This does not interfere with the regular handset and the phone can continue to be used as a standard phone.

Various types of headsets are available. One is equipped with a microphone and an earpiece. Some have headbands, and others fit over the ear and are supported by it.

Memory equipped telephones are well suited for the physically handicapped. Instead of dialing or pushing a series of buttons, just a single button is used for dialing. If the telephone isn't memory equipped, an auxiliary memory dialer can be used.

Single-Button Telephone

Some disabled persons can use the telephone dial but cannot hold a handset. An adjustable arm is available that will hold the handset, as shown in Fig. 3-38. The phone is permanently mounted on the extension arm with an on/off switch on the phone itself. To place or receive a call, the switch on the phone must be turned on. Information about phone extension arms can be had from your nearest city, county, or state rehabilitation center.

Electronic Larynx

An electronic larynx has been developed by the Bell System for people who have lost the use of their vocal cords. It substitutes electronically controlled vibrations for the natural vibrations of the vocal cords. Sound is produced when the instrument is placed against the throat. The user can learn to speak conversationally formed words using the tongue, lips, and teeth in the usual manner. The pitch of the sound is adjustable for men or women.

TELEPHONE FEATURES

The number of possible telephone features is astonishing, but no one telephone can be said to have them all. Some of these features have been touched on in the preceding descriptions of various tele-

Fig. 3-38. Extension arms holds handset. *(Courtesy Bell Systems)*

phone types. They are included here, plus others, so as to supply a comprehensive overview.

You can get a basic telephone or you can get one that can be turned off, that rings a chime instead of a bell, that will dial one number for you, or which you can program to call a long list of numbers. You can have a telephone system that will call you to let you know it has received messages. You can have a telephone that will let you handle an emergency call for help, quietly and quickly. You can have a telephone that will accompany you wherever you may wish to go. You can have a telephone that will display the number you called, or which will announce the numbers as you dial them. You can have a telephone system that will turn on the heat or air conditioning in your home, or will turn kitchen appliances on or off. You can have a telephone which will be your first line of defense in a home security system. And these aren't all of the possible features.

But while a telephone may have a large number of features, these also increase the price of the unit, and also mean more expensive repairs should the unit ever require them. There is not much point in paying for features of you have no use for them, intriguing though they may be. Remember, also, that manufacturers of telephones add features to make their products more competitive.

The fact that you own or are planning to buy one or more telephones that do not have all the features you are interested in doesn't mean you must forego these features forever. Accessories are now available for updating telephones. Thus, you can start with a simple telephone system and add accessories as you wish.

Redialing

This feature is now quite commonly used on telephones. If after dialing a number, you hear a busy signal, the usual approach is to hang up the receiver, wait for a short time and then to try the number again. With a redialing feature your telephone takes over this job and will automatically redial the last number you tried to reach manually. Your phone will ring when the connection is finally made. However, during redialing time your phone will sound a busy signal to anyone trying to reach you.

Repertory or Memory Dialing

Some telephones are equipped with a memory that will let you store numbers you call frequently or emergency numbers, such as police, fire, or hospital. The stored numbers can be dialed quickly simply by dialing the correct access number. It is easy to remember an access number, or you can have these posted adjacent to or directly on the telephone body itself. Some telephones have a memory capacity for 64, 16-digit numbers.

Not all memories have equal capabilities. They are limited in two ways: the total amount of telephone numbers they can store, and in the total digits for each. Some telephones include a battery backup so stored numbers won't be lost in the event of a power failure or if the unit is moved.

Pulse/Tone Switch

A telephone could be equipped with a pulse/tone switch to make it compatible with either rotary or true tone dialing systems for pushbutton convenience.

Electronic Tone Ringer

Some people find the sound of a telephone harsh, unpleasant, and sometimes startling, particularly in the home where telephone ringing can produce a business like atmosphere. An electronic tone ringer supplies a pleasant warbling sound. Its volume, like that of a telephone bell, can be adjusted manually, or can be turned off.

Speakers

A speaker telephone eliminates or minimizes the need for manually handling the receiver. It permits hands-free operation and also allows more than one person to talk on the telephone at one time. If, for example, two members of a family want to use the telephone as a party line, it is not necessary with a speakerphone, for one of them to use an extension phone. Speakerphones can not only be used for family calling but for conference calls as well.

On-Hook Dialing

Speakerphones also feature on-hook dialing. Once dialed, the ring, answering voice, or busy signal is heard through the unit's built-in speaker. Two-way conversation is carried on from anywhere in the room, or, for privacy, once the receiver is lifted, the speaker can be switched off by depressing a mute button when privacy is wanted. The speaker is sometimes referred to as *a monitor.*

Backspace Erase

It is possible to make an error while entering a telephone number into a unit's memory. With a backspace erase feature, you will not need to reprogram the entire number, but just make the needed correction.

Cancel

Sometimes, during dialing, you may change your mind and decide to call another number instead. With the cancel feature you can instantly stop any operation in progress. This eliminates the need for hanging up, and then waiting for a dial tone. The numbers you have dialed are canceled, and you can go right into the procedure of calling the new number.

Time Pause Control

In dialing an access number you must wait for a small amount of time before you follow the access code with the number you care calling. In using an access code, you will note a change in tone. Thus, if you are using digit 1 as an access code for a long distance line lifting the receiver produces a certain dial tone. This dial tone changes when you depress digit 1. After a small amount of time following digit 1 you will hear a new tone.

There is a definite time lapse between the time you depress digit 1 and then the telephone number you are calling. If you do not allow enough time to pass, the number you are calling will not be reached. To avoid this problem, some phones are equipped with progammable time pauses. This means the telephone will automatically allow the correct time lapse.

LED Reference Readout

You may have one or more telephone numbers programmed which you can reach simply by depressing or utilizing a particular button on the telephone pad. An LED Reference Readout will show the telephone number you are calling, that is, it will display the number in the memory, even though you are using just one digit or symbol for making the call.

Long Distance Lines

Alternative long distance lines have their own access codes and these must be used immediately prior to making a long distance call. These access codes are 7 digits and so they do make single pushbutton operation a chore. Some telephones will let you put these access codes in memory, so that dialing when using these services (such as those supplied by MCI and Sprint) will be simplified. And they will be further simplified if you can program in your account number.

Auto Disconnect

When making a telephone call, the ringer of the called party will continue until you hang up. But when should you do this? After three rings? Four? Ten? This means you will need to hold the telephone and count the number of rings, something you may not start doing in the expectation of having your call answered at once. With an auto disconnect feature, the ringing action will stop after six rings, will do so automatically, and your telephone will turn itself off.

Hold Button

With the use of a hold button you can answer one telephone and transfer the incoming call to an extension. The hold button will be released automatically after hanging up without the need for running back and forth, lifting and replacing receivers.

Timer

A telephone timer feature times all outgoing calls automatically. This is an excellent concept if you want to keep track of your telephone costs.

Memory Phone and Clock Radio Combination

The component shown in Fig. 3-39 is a memory telephone with all the conveniences of a clock radio. The 12-number phone memory includes 3 one-touch emergency numbers, recessed to prevent accidental dialing, and memory for nine additional frequently called numbers. Another feature is a phone number index in the unit's base for easy reference of numbers stored in memory. The clock radio has electronic pushbuttons for time and alarm settings, battery back up in case of power failure and automatic radio muting when the phone is in use.

Fig. 3-39. Memory telephone and clock radio combination. *(Courtesy General Electric)*

TELEPHONES OF THE FUTURE

Telephone manufacturers are engaged in two activities. The first is to make their telephones as feature equipped as possible. The other is to design telephones with new applications. Here are some being worked on by the Bell System.

TELESYSTEM CONSOLE

This telephone is supplied with a display system that shows a telephone number in the process of being dialed. This visual display improves dialing accuracy. It not only shows the number being dialed but the time, the day, and the date. Stored numbers can be dialed at the touch of a button.

This telephone will have a provision for the addition of a cartridge and/or modules. In doing so additional services can be provided, such as a personal electronics directory, or an electronic data book. It will also be speakerphone equipped and will supply intercom paging.

TELETERMINAL

The teleterminal combines the power of a computer terminal and a telephone. It can be easily instructed to provide a variety of personal information and communications services: i.e., it can store and display business data, a personal calendar and customized telephone directories. Typewritten messages can be sent or stored and received when desired.

CHAPTER 4

TELEPHONE ACCESSORIES AND ADD-ON DEVICES

Accessories can be put into two categories. The first of these would include items such as plugs, wire, adapters, couplers, etc. The other could be called "add-ons," devices or units that supply an existing telephone with new features. A representative add-on would be a unit that converts a rotary phone to a pushbutton type. It could be a separate ringer amplifier for those who are hard of hearing. A telephone answering machine could be regarded as an add-on for it augments the functions of a telephone.

At one time, when local telephone utilities had the sole right to rent and install telephones and telephone wiring, there were extremely few accessories to be had. Today, with a rapidly growing do-it-yourself phone installation concept, there is an increasing availability of accessories of all kinds, including modular adapter plugs, dual in-line couplers, replacement and extension cords, plus jacks for wall mounted phones. One company has a telephone accessory line that includes 162 products.

The trend in telephones is toward a component setup, an arrangement in which the telephone is the central unit, but has one or more add-ons. The reason for this is that new features are constantly being developed, but no one telephone will have them all. However, it is more economical to buy an integrated telephone, a unit having all the features you want. In the future some add-ons will merge with telephones, resulting in a greater supply of integrated units.

TELEPHONE ACCESSORIES

Buying your own telephone doesn't always mean you can plug it into a jack. The jack (or jacks) may not be suitable, the phone cord may not be long enough, or you may want to install a phone in a room not equipped with jacks. You may want to attach a headrest to your phone to let you use it hands-free, or, if the phone is a rotary dialer, equip it with a pulse tone attachment.

Many telephone accessories can be categorized as modular or standard. Modular accessories are for the more modern type of phone.

Modular Telephone Wire

This consists of wire (Fig. 4-1) equipped with a modular plug on one end and four color coded spade lugs on the other. You can use modular telephone wire to change standard telephones, standard jacks and answering devices to modular form.

Modular Coil Extension Cord

This is coiled modular wire (Fig. 4-2) with modular snap-in, snap-out plugs on both ends. You can use this type of cord to extend

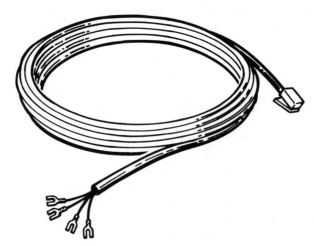

Fig. 4-1. Modular telephone wire. This is a straight cord with a modular plug on one end and four spade lugs on the other. The modular plug fits into a corresponding wall jack, and the spade lugs are connected to the telephone.
(Courtesy Dur-O-Peg)

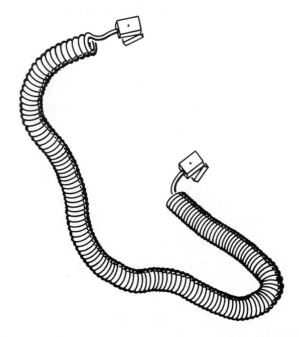

Fig. 4-2. Modular coil extension cord equipped with modular snapout plugs on both ends. Also available as a straight line extension. *(Courtesy Dur-O-Peg)*

modular phones or phone handsets. These cords are available in various lengths, up to 25 feet, with colors in yellow, blue, gold, white, and ivory. To get a good color match take along your old cord or else buy the cord in the same store in which you bought your telephone.

Modular Jack and Wallplate

A jack for a telephone can be compared to an ac outlet. Fig. 4-3 shows a wallplate for use with a modular jack. The plate, and its accompanying modular jack can be flush mounted and used to convert a conventional or standard jack installation to a modular system. Unlike an ac outlet, however, there is no danger from voltage.

Modular Jack Adapter

This accessory (Fig. 4-4) consists of modular wire with a modular plug at one end and a standard jack on the other. It converts from a standard jack system to modular and can be used with standard telephones and answering machines.

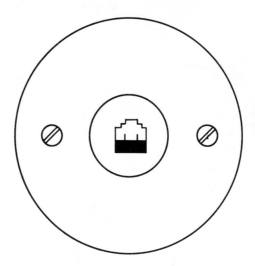

Fig. 4-3. Modular jack and wallplate *(Courtesy Dur-O-Peg)*

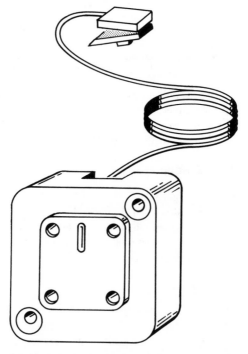

Fig. 4-4. Modular jack adapter. *(Courtesy Dur-O-Peg)*

CORDS

A cord is a length of wire covered by an insulating material. Cords are generally available in standard lengths such as 6, 12, 15, 20, and 25 feet, but you can purchase any length you want for on-premises wiring.

Cords are used to connect a handset to the telephone base. These are generally tightly coiled so as to have the minimum visible amount of cord, but to let you walk some distance away from the base of the telephone while holding the handset in your hands. The coiled length can range from about 1 foot to 4 feet; the extended length up to about 25 feet. Since the coiled cord will be highly visible, you should get a cord whose color matches that of your telephone.

The other type of cord is a straight-line type, used for on-premises wiring, possibly for installing a new telephone outlet. These cords may come equipped with spade terminals on both ends and in that case the terminals should be color coded for easy identification. The cords may also be equipped with an accessory on one end or on both ends, accessories such as modular or standard jacks or plugs.

Straight line cords can be snaked through walls, just as you would with wire for ac use. An easier way is to staple the cord along the baseboard. It is less work but it is also less attractive.

Modular Extension Cord and Duplex Jack

You can connect two telehones to the same telephone outlet by using the component shown in Fig. 4-5. It consists of a 25-foot extension cord, a modular plug at one end and a twin modular jack at the other. The two plugs of the telephones are inserted into the twin jack arrangement. At the opposite end the modular plug is put into the modular jack in the telephone wall plate.

Using this modular extension cord, both connected telephones will ring at the same time and either phone can be used for making out-

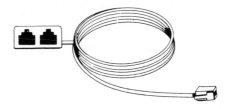

Fig. 4-5. Extension Cord with dual modular jacks. *(Courtesy Recoton Corp.)*

side telephone calls. When a telephone call is received, one of the phones can be used for answering and the other as an extension phone.

Universal Combination Jack

You may have two telephones in your home, one of which has a standard plug and the other a modular plug. This does not mean you need separate jacks for each. Instead, you can use a universal combination jack, as shown in Fig. 4-6. The jack will simultaneously accommodate a standard plug and a modular plug. No wiring is necessary and all connections are made automatically. The device can be plugged into an existing four prong jack or easily wired to a junction box.

Modular Adapter

Fig. 4-7 shows a standard four prong telephone plug, but it is somewhat different from the usual type. It contains a recess, shown on the front of the unit, to accommodate a modular plug. The newer telephones are all modular equipped. To use this accessory, put the modular plug of your telephone into the adapter. Then insert the four prongs into the standard wallplate jack, generally positioned near the baseboard in the room in which you have your telephone. No wiring changes are needed. Nor is it necessary for you to change the four-prong standard wallplate you are now using.

Modular adapters can be one piece as shown in Fig. 4-7 or two-piece types as illustrated in Fig. 4-8. Aside from the physical differences they are identical in the way they work. As indicated in Fig. 4-8 the line-cord plug is inserted into the modular converter and that part, in turn, is plugged into the wall-mounted receptacle. As with the single-piece accessory, no wiring changes are required.

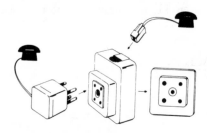

Fig. 4-6. Universal combination jack. *(Courtesy Recoton Corp.)*

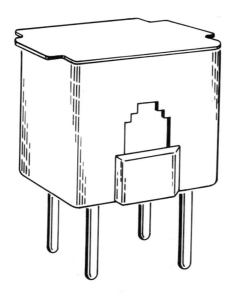

Fig. 4-7. Modular adapter. *(Courtesy Dur-O-Peg)*

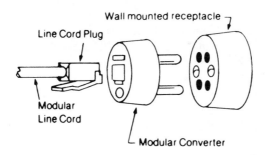

Fig. 4-8. Two-piece modular converter. *(Courtesy Comvu Corp.)*

Modular Jack

The modular telephone jack shown in Fig. 4-9 is intended for surface mounting. Use it for installing a new modular system or converting a standard jack to modular.

The advantage of a surface type modular jack such as the one in Fig. 4-9 is that it is so easy to mount. All you need to do is to attach it to the baseboard with a pair of machine screws. However, it does not supply a finished appearance.

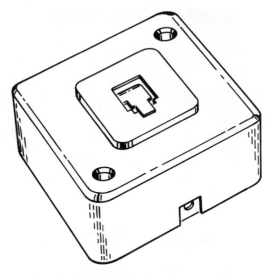

Fig. 4-9. Modular jack. *(Courtesy Dur-O-Peg)*

The flush wall mount unit shown in Fig. 4-10 supplies a more professional look, but it means the jack will need to be recessed in the wall. This is no problem if you own your own home but if you are renting an apartment the terms of your lease may prohibit you from doing so. Fig. 4-11 shows the wiring of a modular jack.

As shown in Fig. 4-12, you can connect a modular telephone to its jack in two ways, depending on the type of wall jack you have. If the jack is a modular type (drawing at the left) and the phone cord is equipped with a modular plug, just insert the plug into the jack. If the jack is a standard type as shown in the drawing at the right, just interpose an adapter between the plug and the jack.

STANDARD TELEPHONE ACCESSORIES

There are two ways of handling a telephone that just does not seem to let you get far enough away from the telephone jack. One method is by using a longer phone cord. Those now available have tight coils and so the cord, when not pulled, is much shorter than in its extended form.

Another way of handling this problem is to move the phone outlet to a more convenient place. This can be done with a telephone ex-

Fig. 4-10. Telephone wall mount with modular jack *(Courtesy GC Electronics)*

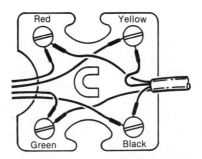

Fig. 4-11. Wiring of a modular jack. The red and green wires connect to the telephone line and are for your basic service. *(Courtesy Mitel Corp.)*

tension cord of the kind shown in Fig. 4-13. It consists of a telephone cord, generally 25 feet long, with a standard telephone plug at one end and a standard jack on the other. Simply plug the standard jack into the phone outlet and the telephone into the four-terminal jack on the other end.

TELEPHONE CORD LENGTH

Telephone cords vary in length, but typically are about 10 to 14 feet. However, this is the fully stretched out length, and, in effect lets

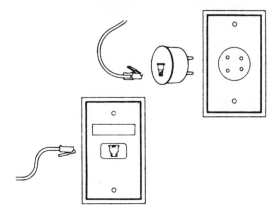

Fig. 4-12. Two ways of connecting a modular telephone. The modular plug connected to the telephone cord fits into a four-prong adapter. The adapter can then be inserted into a standard jack. The lower drawing shows a phone cord equipped with a modular plug. This can be inserted into a modular jack which is mounted on a wall plate. *(Courtesy GTE Consumer Sales Div.)*

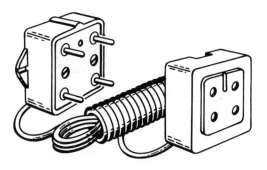

Fig. 4-13. Standard telephone extension cord.
(Courtesy Dur-O-Peg)

you move as much as 14 feet, maximum, from the telephone. A cord of this kind when coiled should retract to about 4 feet. Thus, two numbers are important when considering a telephone — the maximum cord length and the retracted length. Ideally, maximum cord length should be as much as possible, but within practical limits. A stretched length of 14 feet should certainly be enough. The retracted length should be as short as possible.

Sometimes a telephone cord is inadequate. This could be either a too-short length between the telephone and the wall connecting plate or between the telephone receiver and the phone, or both.

Either or both cords can be replaced by a new cord that extends to as much as 25 feet and retracts to as little as 4 feet. These replacement cords are available in a variety of colors including white, ivory, beige, yellow, green, red, blue, brown, and black. They will fit all telephones using clip-in cords in jack-equipped homes or offices. They can be clipped into position in seconds, but are intended only for modular units.

If your phone system uses standard connectors instead of modular you can use standard telephone wire, as shown in Fig. 4-14, for extending the practical operating distance. The wire is equipped with color-coded lugs on each end.

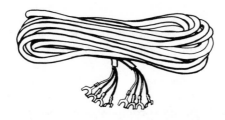

Fig. 4-14. Standard telephone extension cord is flexible and straight. It is equipped with color coded terminals on each end and is used for connecting standard phones and wall outlets.

SILENCER CORD

Appliances and electronic devices such as radio receivers, tv sets, electric fans, and hi/fi systems are equipped with a switch to enable the user to turn power on and off. A telephone, though, is not generally regarded as being power operated, and yet it is, with the telephone utility supplying power for the ringing circuits. Delivery of the ringing current is not controlled by the owner of the telephone but by the telephone company.

This raises a problem since there may be times when you do not want the sound produced by the ringer. There are some phones equipped with an on/off switch that has a triple function: off, low, and loud. One alternative is to disconnect the phone plug from its outlet, whether modular or standard. This is not a recommended action since the connectors are not designed for such usage and could be damaged. An alternative is to use a silencer cord. It stops your phone from ringing when you don't want to be disturbed. The cord replaces any cord that connects the phone to the wall jack, as long as the cord has modular plugs on both ends.

The silencer cord has a handy switch that keeps your phone from ringing. Callers hear the phone ring, but you do not. The component meets all FCC requirements for customer installation. You can use this cord on any phone whose ringing you want to control.

The unit also has a reminder feature that alerts you to the fact your telephone is off. When you try to make a call you only get silence instead of a dial tone. Since the phone is actually disconnected you can also turn your phone's handset upside down or take it off the hook as a visual reminder.

T Adapter

You may sometimes want to have two telephones in the same room, possibly adjacent to each other on a desk. However, you may have only one telephone outlet in that room. Instead of going to the expense of having another outlet installed, you can use the T adapter shown in Fig. 4-15. One side of the unit plugs into the existing modular wall jack. The front of the component will accept a pair of modular plugs, supplied by the two telephones.

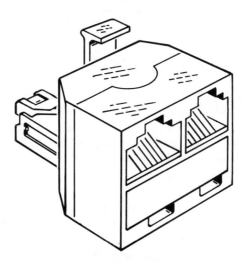

Fig. 4-15. T-adapter can be used to connect a pair of telephones to a single telephone outlet. (Courtesy Zoom Telephonics)

Cord Coupler

You can connect a pair of modular cords by using the accessory shown in Fig. 4-16. The clip arrangement on the plugs of the modular cords is such that it will not come loose when inserted into the cord coupler.

Standard Telephone Plug

Most of the older-type telephones are equipped with a four-prong telephone plug of the type shown in Fig. 4-17. These plugs are intended for use with standard three or four conductor telephone wire. The extension from the plug at the left center of the unit is a plastic ear. Its purpose is to help you remove the plug after it has been inserted into its jack, if you should need to do so.

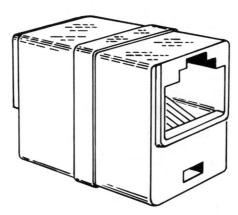

Fig. 4-16. Cord coupler can be used to connect a pair of modular telephone cords. *(Courtesy Zoom Telephonics)*

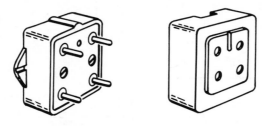

Fig. 4-17. Standard telephone plug (left) and standard telephone jack. *(Courtesy Dur-O-Peg)*

Standard Telephone Jack

A standard telephone jack is a receptacle having four holes to accommodate the insertion of a standard telephone plug. The jack shown in Fig. 4-17 is a surface mount type and is intended for use with three or four wire telephone cord.

Standard Jack and Wallplate

Fig. 4-18 shows the hardware necessary for mounting a standard telephone jack such as the type previously illustrated in Fig. 4-17. In the drawing in Fig. 4-18 the standard jack is shown once again and directly below it are the two bits of mounting hardware. The one at the lower left is the holding plate for the jack. It fits over the jack. The holding plate has a pair of "ears," one above the plate, the other below it. These are fastened into the wall by screws.

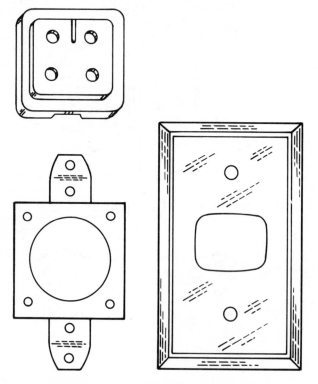

Fig. 4-18. Standard jack (upper left), holding plate (lower left) and finishing plate (lower right). *(Courtesy Dur-O-Peg)*

The part shown at the right in the same drawing is a finishing plate. After the holding plate has been screwed into the wall and is holding the telephone jack in position, the finishing plate is slipped over the holding plate. It is held in place by a pair of machine screws. These fit into the two holes shown in the drawing. The idea is similar to that used for ac power outlets.

HOW TO CONNECT A TELEPHONE

It is easy to connect a telephone to a phone outlet. The phone should be equipped with a length of wire terminating in a plug that is either the so-called standard or modular type. The standard plug has four pins. Simply insert the plug into its corresponding jack. There is no danger of shock.

The procedure is even easier for phones equipped with modular plugs. The wallplate may have its modular jack near the bottom part of the wallplate, as shown at the left in Fig. 4-19 or it may be at the center of the plate as in the drawing at the right. Insert the modular plug until you hear a click. The plug should fit in easily and enter smoothly.

Modular plugs are equipped with a squeeze tab. This tab produces a slight clicking sound when you insert the plug into the jack. This sound is your assurance that you have inserted the plug correctly into its jack.

The purpose of the squeeze tab shown in Fig. 4-20 is not only to let you know you have inserted the plug correctly, but to hold the plug in place. To remove the plug you will need to push up on the bottom

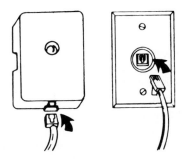

Fig. 4-19. Modular wall plate may have jack at bottom (left) or center (right) of the wall plate. *(Courtesy Comvu Corp.)*

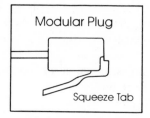

Fig. 4-20. Squeeze tab helps hold modular plug in its jack.
(Courtesy Zoom Telephonics, Inc.)

part of the squeeze tab. Do not try to force or pull the modular plug out unless you do this.

WHAT IF YOU HAVE NO JACK?

It is possible that the room in which you want a telephone is not equipped with either a standard or a modular jack. In that case you have two options. You can contact your local telephone utility and ask them to install a jack. Or, you can buy your own and install it yourself. Some dedicated telephone stores also have an in-home on-premises wiring service.

Your present telephone may also be connected to a small junction box. This is known as a wired telephone. If your telephone is wired into a wall outlet call your local telephone company and ask that a USOC RJ11C connector be installed.

TELEPHONE IN THE BATHROOM?

Water and electricity make a dangerous combination. The voltage delivered to your telephone is relatively small and the current availability is limited, so there is no danger of shock. However, as a general safety precaution it is advisable to keep any and all electrical appliances out of the bathroom. Some of the add-on units for telephones are ac powered and these are highly dangerous for bathroom use. The problem is that a telephone in the bathroom can produce a false sense of security and so the add-on unit may accompany it without much thought.

ADD-ON UNITS

All of the components described so far are accessories. There are actually hundreds of these so if there is a particular unit you want do not assume it isn't available just because it has not been described here. It may be difficult to find a supplier, so your best option is to visit or write to a dedicated telephone store or possibly the manufacturer of your telephone. Most required accessories, though, will be either wire, plugs, or jacks.

Like accessories, you will find a large number and variety of add-on units. The purpose in buying an add-on unit is to supply your telephone with some feature or features you consider desirable.

Before buying an add-on consider whether you own your present telephone or rent it from your local utility. If you rent, you will be buying an add-on to use with someone else's property. If you own the telephone, it may be that you can buy an integrated telephone (Fig. 4-21) that has all the features you want. There are two advantages in doing so. The new unit may require less room than your old phone plus an add-on. And, since the new unit is integrated, you are assured that the telephone and add-on are designed to work together.

Fig. 4-21. Integrated telephone is equipped with a battery-powered memory.
(Courtesy Panasonic Consumer Electronics Group)

ROTARY TO PUSHBUTTON DIALING

You can convert from rotary to pushbutton dialing in several ways. If you have a rotary dial phone you can change to pushbutton by installing a new telephone, checking with your phone company first to make sure their lines are tone adaptable.

As an alternative, if you do not want to give up your rotary dial telephone, you can get pushbutton dialing units that will fit into your present rotary telephone. One such device is shown in Fig. 4-22. It is equipped with the same pad as the average pushbutton unit. To use

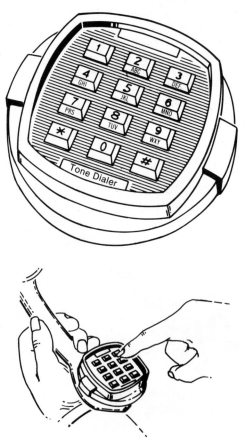

Fig. 4-22. To use touchbutton dialer, remove telephone's handset mouthpiece and screw unit in its place. The dialer contains a microphone positioned between the keys.

this component all you need do is to remove your telephone handset's mouthpiece (this part contains the microphone) and screw on the accessory in its place. You now have the advantage of being able to dial directly from the handset. This accessory is equipped with a microphone. The keys of the telephone pad have a raised outer ridge to make more accurate dialing possible.

ADD-ON DIALERS

The unit shown in Fig. 4-23 converts rotary dial phones to pushbutton, making that phone easier and faster to use. It stores 10 frequently called numbers and will dial any of them at your command. It dials accurately and helps end wrong-number dialing. It can store up to 16 digits per memory location and can be used to dial anywhere in the world. It operates by using a 9-volt alkaline battery that should last for more than two years. No external power supply is needed. It plugs into single line phones via modular jacks.

This dialer does not use pushbuttons. Instead it has a touch-

Fig. 4-23. Add-on dialer converts rotary dialing to touch sensitive dialing.
(Courtesy Dictograph U.S.A.)

sensitive pad for dialing and uses the touch-sensitive technique for memory calling. As shown in the photo, it can be positioned conveniently directly below the rotary dial telephone. It can also be used with pushbutton telephones to add memory capacity. The unit also has a last number redial feature.

Portable Tone Dialer

You can carry a tone dialer with you and use it with any telephone. No connections or wiring are necessary. All you do is hold the portable dialer (Fig. 4-24) against the handset mouthpiece to make an acoustical connection to the phone line. The component is battery operated. Models are available with memory capacities of 20, 40, or 80 memory positions. The component, equipped with index aids to make it easy to store and locate numbers, operates with a standard 1.5-volt battery. Since the unit does not interface with the telephone lines it does not require FCC registration.

Telephone dialers run the gamut from simple to highly sophisticated. The unit shown in Fig. 4-25 falls into the latter category for it can store and automatically dial up to 176 phone numbers, but also has one-button redial. It "breaks through" to busy numbers with repeated rapid redialing. It will repeatedly recall unanswered phones, when the called number doesn't reply, and automatically calls back

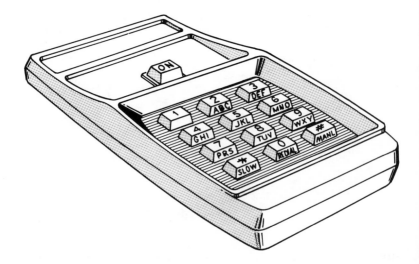

Fig. 4-24. Portable pushbutton dialer.

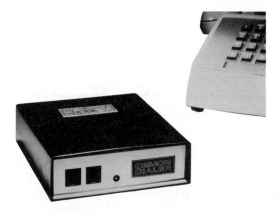

Fig. 4-25. Add-on unit supplies telephone with numerous sophisticated capabilities. *(Courtesy Zoom Telephonics, Inc.)*

every 10 minutes for up to 10 hours. It will dial numbers up to 32 digits long, making it ideal for Sprint, MCI, and other alternate long distance phone systems. It supplies true tone recognition and will detect dial and computer access tones. You can use it as an add-on for your existing phone system, whether tone or rotary dialing types, single or multiline.

ADVANTAGE OF MEMORY

The advantage of a memory bank is that it lets you use just a single button on the telephone for making a call. It eliminates calling a wrong number and it permits much faster calling. If, for example, a telephone has a 10-digit memory bank, every one of the pushbuttons on the pad can be used for that memory. The memory bank, as indicated in Fig. 4-26, could consist of a physician, hospital, police station, office, school, personal friends, and relatives.

Memory Stringing

This feature, mentioned briefly earlier, is the ability of a telephone to operate its memory bank sequentially. Thus, a telephone could have a single 10-digit storage capacity. This means the phone could store any number up to 10 digits and these 10 digits could be called out by depressing just a single button. Some telephones have a mem-

1	Bank	Paul	6
2	Police	Mother	7
3	Fire	Ken	8
4	School	Alex	9
5	Office	Jerry	0

Fig. 4-26. One possible arrangement of a memory bank using all 10 digits of a telephone pad.

ory with a much larger storage capacity. And some permit memory stringing. With memory stringing memories function one after the other. Thus, if a phone has two 10-digit memory banks you could store as many as 20 digits with memory stringing.

Memory Failure

Telephone memories require a source of power. This can be supplied by the ac power line or by batteries. Both have advantages and disadvantages. In the event of power-line failure, memory is lost, but it can be returned when power is restored by punching in new data.

The memory system can also be battery operated, but these do not last indefinitely. The best arrangement is an ac-powered memory that uses a battery backup system in the event of a power failure. However, even with this procedure the batteries should be checked from time to time. The battery types are generally AA units. The unit shown in Fig. 4-21 has a battery-powered memory.

TELEPHONE RINGER

While your telephone probably has a detent type volume control, generally in the form of a slide switch, it is always possible to miss a telephone call if the phone is located in a remote room, or if the ambient noise level is high, or if the persons having the telephone have a hearing loss.

You can get an add-on ringer which is connected across the telephone line. When the telephone rings, so does the accessory ringer. If the ringer is placed in a hall in the home, the telephone can be heard for quite a distance.

Ringers can be battery-operated types, often using four NiCad cells. The length of operating time of the batteries depends on how often the telephone rings, but the batteries should be good for about a thousand calls. Some ringers are equipped with builtin battery chargers, minimizing the battery replacement problem. Other ringers are fully ac operated and in that case will work indefinitely.

Not all ringers are alike any more than all automobiles are alike. When buying a ringer get one that has as high an input impedance as possible. The higher this impedance the smaller the amount of telephone ringing current the unit will need. This means you can install the unit and it may not be detected by telephone utility monitoring circuits.

It is, however, a matter of your own conscience. The telephone lines do belong to your telephone utility and any energy taken from those lines should be paid for. Further, it is also a matter of legality. Any device connected across telephone lines utilizing current does need to be reported to the telephone utility, and you should supply them with the telephone ringer equivalency number. Any ringer you buy should be FCC type approved.

HOLD CONTROL

Most telephones intended for in-home use are not equipped with a hold-call feature. You can get an add-on unit that will eliminate the inconvenience of covering the mouthpiece when talking on the phone. The component, in the shape of a small cylinder, is inserted between the handset and the base of the phone, and, since it uses modular connectors, is easy to install. It can put any call on hold simply by depressing a button on the unit. No wiring changes are required inside the phone.

TELEPHONE AMPLIFIER

There are now available telephones equipped with a builtin monitor speaker so you have the option of using the handset or the speaker. But if your telephone does not have this feature you can equip it with the telephone amplifier arrangement shown in Fig. 4-27. It consists of two separate parts: a pickup and an amplifier. The pickup comprises a rubber suction cup at one end of a connecting cord, and a jack plug at the other. The cup is moistened and fastened to the telephone, somewhere near the mouthpiece. When the telephone is used the amplifier is turned on and both sides of the conversation can be heard through the speaker. This is not a recording device, so you do not need to notify your callers that the amplifier is being used. And, since it does not interfere with the telephone lines you need not notify the telephone company either.

SOFT-TONE ADD ON

One of the advantages of owning your own telephone, other than the reduction in your monthly telephone bill, is that it is your property and you can make any changes in it you wish, or any additions provided you do nothing to interfere with or upset telephone line currents.

Not everyone has the same sense of hearing and some are offended or are sometimes startled by the harsh, loud ringing of a telephone bell. Your telephone may come equipped with a switching type of volume control which will let you adjust the sound level of the ringer, but generally only within rather coarse limits.

You can replace the existing telephone ringer with a soft-tone add on. You can also use this device as a supplementary warning ringer in another room. As shown in the chart given below, add-on ringers which deliver a soft musical tone can be had that supply tones with a varying degree of loudness.

LOUDNESS	SOUND LEVEL	FREQUENCY
soft	70 dB	3500 Hz
medium	80 dB	2900 Hz
loud	95 dB	2900 Hz

The two tones having frequencies of 2900 Hz will sound a bit deeper than the tone frequency of 3500 Hz. The difference in sound level from soft to medium and from medium to loud is quite substantial.

The existing ringer in your telephone is connected across a pair of wires color coded green and red. To locate the ringer (your telephone may have two of them), open the telephone case and find the ringer and the wires going to this part. The easiest way to connect the add-on is to disconnect just one of the wires of the ringer — it isn't

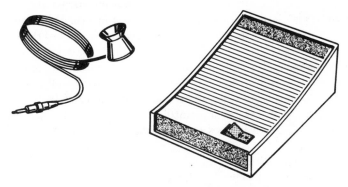

Fig. 4-27. Telephone amplifier. *(Courtesy GC Electronics)*

necessary to disconnect both — and connect the two wires of the add on to the two wires leading to the ringer.

The ringer wires are not polarized. This means either wire of the add-on ringer can be connected to the red or green wires in your telephone.

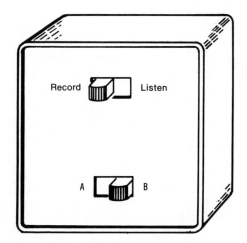

Fig. 4-28. This component, plus an audio tape deck, will record all your phone calls automatically. *(Courtesy The Fone Booth)*

PHONE CALL RECORDER

You can record all of your incoming phone calls by using a TAD — a telephone answering device, such as any one of those described later in Chapter 7. However, you can also use the add-on unit shown in Fig. 4-28 provided you have a tape recorder with a remote input. This device will automatically record all your calls, will start taping telephone conversations when the telephone is in use and will stop when the call is completed. The device is solid-state, requires no batteries or ac power. Since it does not interface with the telephone company's lines there is no additional charge added to your bill, nor is it necessary for you to inform the phone company of its use.

One of the disadvantages of this component is that it is a record-only unit, and, unlike a TAD, does not have an announce tape. Consequently, you should inform the person with whom you are having a conversation that it is being recorded.

TIME AND TEMPERATURE TELEPHONE ADVERTISING SYSTEM

This component answers phone calls with an advertising or promotional message and then supplies the current time and temperature. Consumers initiate contact by dialing your dedicated phone number for the time and temperature. It provides a 24-hour source of time and temperature information to your local community, and counts every caller who hears your message daily. The component consists of a commercial telephone answering system interfaced with a microprocessor system which features a specially developed computer voice that announces the current time and temperature over the telephone lines. The unit has the ability to answer thousands of calls per day with one phone line. It operates with a standard 120-volt ac line and plugs into a phone jack just like an ordinary telephone. In case of a power failure, it has backup power for up to 5 hours of continuous operation.

Technically, this should not be termed an add-on, for it is really a dedicated phone answering unit. But its function is that of supplementing your existing telephones which can be used for regular telecommunications.

CHAPTER 5

CORDLESS TELEPHONES

Telephones can be categorized in many ways. One is to list them as either corded or as cordless. A corded phone is one connected directly to the telephone lines by pairs of wires sometimes called a cable.

A cordless telephone (Fig. 5-1) also known as a cord-free or wireless telephone consists, in its basic form, of two units. One of these is the base (or transponder) and is located near a corded telephone. In some arrangements the base unit and the corded telephone are separate items; in others, the base unit and the telephone may be part of a single assembly, integrated and mounted on a common base.

The second part of the cordless telephone is the hand unit, sometimes called a handset, a remote handset, or simply a remote. Communications between the base and hand units is duplex; that is, the setup permits two-way conversation just as you have with a telephone.

Both units, the base and handset, are radio-operated devices. Both have circuits for generating an fm (frequency-modulated) signal and both work very much like miniature fm broadcast stations. Thus, the base unit is equipped with an fm transmitter and receiver, and so is the hand unit.

CONNECTING THE BASE STATION

Fig. 5-2 shows the two easy steps to follow in connecting the base station. Insert the male ac plug of the base unit into the nearest avail-

171

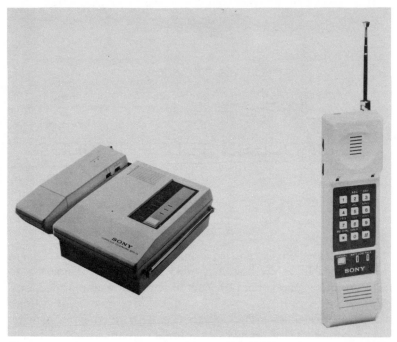

Fig. 5-1. Cordless telephone base station and handset (left). The photo at the right shows the cordless telephone handset by itself.
(Courtesy Sony Consumer Products)

able ac electric power outlet. Then connect the modular plug of the base unit into a telephone receptacle. If that is already occupied by the telephone plug of your phone you can get an accessory that will let you use that single telephone outlet for both your telephone and the base unit. Finally, extend the telescoping antenna to its full length.

HOW THE CORDLESS TELEPHONE SYSTEM WORKS

When the base unit receives either a voice or a code signal from the telephone, assuming someone is calling in, the signal modulates a radio-frequency carrier wave, transmitting it via your house wiring as a carrier current antenna. This modulated radio signal, traveling at the speed of light, reaches the telescoping antenna of the hand unit. Inside the hand unit, a fixed-frequency radio receiver detects the calling signal, demodulates it and feeds it into a builtin speaker.

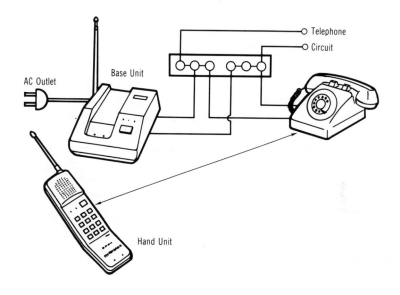

Fig. 5-2. Circuit arrangement of a cordless telephone. Duplex operation is possible between the hand unit and the base unit. Voice transmission from the hand unit reaches the base unit via fm radio transmission. The base unit demodulates or extracts the voice signal from the radio wave and feeds the voice signal into the telephone lines. Called number pulses, received from the hand unit, are brought into the telephone circuit.

Like the base unit, the hand unit's transmitter is capable of transmitting a voice signal and a pulse code produced by depressing the keys on its telephone pad. These signals modulate a radio-frequency carrier wave generated by the miniature transmitter inside the hand unit. The modulated wave, received by the antenna of the base unit, is fed into the base unit's receiver and is then demodulated.

Function of the Carrier Wave

The purpose of the carrier wave, a wave having a high frequency, is exactly what its name implies — a carrier. All it does is carry the voice message to and from the base and hand units. The process of loading the voice message onto the carrier is called modulation. Once the message has been delivered the reverse process, demodulation, takes place. In demodulation the voice message is stripped from the carrier.

CORDLESS TYPES

There are two types of cordless units: the cordless-handset telephone, and the cordless telephone.

Cordless Handset Telephone

The cordless-handset telephone is comparable to an extension telephone that can be used for receiving incoming calls, but not for making any outgoing calls. The cordless-handset telephone is not physically connected to the base and it has no dial. It can respond to incoming calls received by the base and you can use the cordless handset telephone this way at any distance within the range of the transmitter in the base unit. Because it does not have a dial, the cordless handset telephone cannot originate phone calls. To dial, you must use the base station.

Cordless Telephone

The remote unit of a cordless telephone (Fig. 5-3) is equipped with a dial and so it can originate outgoing calls. These are transmitted to the base unit via radio. Most cordless systems today are of the cordless telephone type with a full-functioning remote handset.

Fig. 5-3. The advantage of the remote handset is that it can be carried anywhere. *(Courtesy Electra Co., Div. of Masco Corp. of Indiana)*

You can dial a telephone number with the remote unit of a cordless telephone just as you would with any other kind of phone. The generated tones are loaded on the much higher frequency carrier, and then picked up by the receiver in the base station. The pulses, representing the called number, plus the following voice message are not brought into the telephone associated with the base unit, but are sent directly into the telephone lines.

CORDLESS TELEPHONE OPERATING DISTANCE

The operating distance of cordless telephones ranges from a low of about 300 feet to as much as 25 miles. These distances — the distances specified by cordless telephone manufacturers in their advertising and sales literature — assume ideal operating conditions: flat terrain, no interfering or blocking masses of metal such as chain link fences or structures containing metal, no electrical interference, or competition from competing cordless telephone systems.

While cordless telephone manufacturers emphasize operating distance, possibly as a desirable feature, the actual working distance required should be the minimum needed to assure good communications. Having a tremendous operating distance may supply ego satisfaction but is not always desirable. The greater the operating range available, the greater is the possibility of losing message confidentiality. If your home is on a property that measures 75′ × 200′, representing the length and width, then the maximum possible distance would be along a diagonal line representing the extreme separation points and this is only about 213 feet. Allowing a 25-percent margin for terrain and metal objects would bring this figure up to 266 feet. A cordless telephone having a claimed range of 300 feet would do well in this example as far as communications distance is concerned.

A long operating range can have disadvantages. The first of these is the possibility that a pair or possibly several pairs of cordless telephones used by a group of neighbors could result in mutual interference. There is always the additional chance that each could answer the other's telephone calls. The overall effect could easily be somewhat similar to using a party line. Still another disadvantage of using high-power cordless units, for that is what long range cordless transmission means, is there is a possible loss of privacy. Every time you use a cordless phone you lose the confidentiality supplied by a wired line. Cordless communications is "broadcast communications" and that means anyone with a sufficiently sensitive base unit, or an fm receiver tuned to your operating frequency, can listen in on your conversation.

WHY FM IS USED

Fm (frequency modulation) transmission is used instead of am (amplitude modulation) since this form of signal transmission is less susceptible to electrical noise interference. This lets you use electrical noise makers such as fluorescent lights, a vacuum cleaner, a refrigerator, and other appliances in the same room with the base station, free from their interference.

OPERATING POWER

Both units, the in-base telephone (Fig. 5-4) and the remote handset, require operating power. There are three kinds: that supplied by your ac outlets; that furnished by your local telephone company and that obtained from batteries.

AC Power

First consider the in-home setup. This consists of two parts: the base unit and the telephone. The base unit is connected to the ac power line, while the telephone is attached to the telephone lines. The remote handset, not having the convenience of either of these outlets, is battery operated, generally using nickel-cadmium batteries since these are rechargeable. With some units recharging is done simply by putting the handset into its cradle in the base. This is very convenient since the handset should always be brought back into the house anyway.

The Battery

The portable handset is equipped with 4.8-volt nickel cadmium rechargeable batteries. When you receive your cordless system you will probably be supplied with batteries in a fully discharged condition. It will take between 5 to 10 hours to charge the batteries to the point

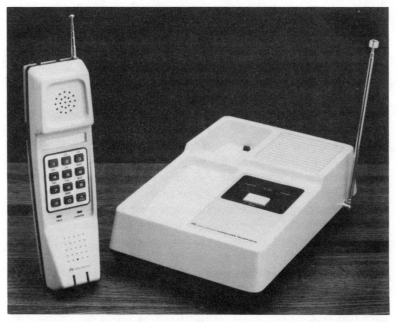

Fig. 5-4. Batteries for the remote handset are automatically recharged when the handset is positioned in the recess provided for it in the base.
(Courtesy Midland International Corp.)

where they will be able to operate the remote handset. To charge the batteries, connect the base unit to an ac outlet and then position the handset in the space provided for it on the base unit (Fig. 5-4). If the base unit is equipped with a charge indicator (generally a light emitting diode) this will begin to glow. Charging is done automatically and there are no charging controls to adjust.

The length of charging time is variable and depends on the existing state of charge of the batteries. As a general rule of thumb you can count on two hours of charging time for each hour of use of the remote handset. As the batteries become charged they develop a voltage that is in opposition to the applied charging voltage. As a result, the amount of charging current gradually becomes less. Consequently, you will not damage the batteries by leaving them in the charger. All that will happen is they will be on "trickle" charge, with a very small amount of current going into the batteries from the charger. A good charging technique is to leave the batteries on charge overnight.

In this way you will be able to fully power the remote handset for daytime use.

CORDLESS-PHONE ANTENNAS

Both the base unit and the remote use telescoping antennas. These should be extended to their maximum lengths before using the cordless system. No damage will be done to the components if you do not, but you may not be able to establish communications between the two units: the in-home base and the remote handset. As mentioned earlier, the base station uses the ac lines as a transmitting antenna, but it also has a telescoping antenna for receiving. The remote uses a telescoping antenna only, for receiving and transmitting.

Two different fm operating frequencies are used. For transmitting, the base unit, via the ac wiring in your home, sends out a signal at a frequency of 1.7 MHz (Megahertz). There is no charge made by your local power utility for this use of their lines for two reasons. You are not taking any ac power from those lines, and wiring you use is in your home and is your property. The frequency of 1.7 MHz, or 1,700,000 cycles per second, is the carrier wave. A voice signal received by the base station frequency-modulates this carrier wave, and the combined waves are then radiated by your ac power lines.

While a pair of operating frequencies have been indicated, not all cordless telephones use these. The manufacturers of such systems have a choice of five channels.

Quite commonly, cordless telephone systems use 1.7/49 MHz for duplex transmission, although you will find a few that are 49 MHz/ 49 MHz operated. There are advantages to both techniques. With a system using 1.7/49 MHz there is less likely opportunity for interference between the two signals. But for a 49 MHz/49 MHz system the antennas can both be smaller and there is decreased opportunity for electrical interference.

Detailed Operating Frequencies

While frequencies such as 1.7 MHz and 49 MHz have been mentioned, these are descriptive only and do not indicate the exact frequencies. There are five channels and while the following listing shows them in a certain order, any frequency combination can be used. The operating channels are designated as 1A, 7A, 13A, 19A and 25A.

Channel	Handset	Base
1A	49.830	1.690
7A	49.845	1.710
13A	49.860	1.730
19A	49.875	1.750
25A	49.890	1.770

The numbers under "handset" and "base" are all in MHz (megahertz). Note that the handset frequencies are separated by 0.015 MHz while the base frequencies are separated by 0.020 MHz.

Manufacturers are allowed to select any of these frequencies they wish. However, any frequency change must be made by the manufacturer of the unit. Thus, if you and a neighbor happen to have cordless units working on identical frequencies, and if these units are long-range types and have no security feature, the best course of action to take is to return the equipment for a frequency change.

Antennas

The base station also has a telescoping antenna each of whose sections can be recessed. This is the receiving antenna and is used to pick up transmitted signals from the handset at a frequency of 49 MHz.

Because they can use the same frequencies it is possible to get interphone interference when cordless telephone systems are used near each other with each having a high transmitting range. The operating frequencies are those specified by the Federal Communications Commission for this purpose and may not be changed by the manufacturer or by the user.

As indicated earlier, a relatively small number of cordless telephones have handsets which can be used for receiving purposes only. This means they can receive a telephone call from the base station and can listen to a message, but cannot answer. Most cordless telephone systems, though, have handsets which can either receive a telephone call or can originate one.

SIMPLEX AND DUPLEX OPERATION

A few cordless phones use simplex, meaning only one person can talk at a time, an arrangement similar to that used in CB radio, with intercoms or with walkie-talkies. Most wireless phones, though, are duplex and so operation is just like a standard telephone.

Connecting the base telephone to the phone lines is just as easy as connecting the usual telephone. The base unit is equipped with a modular plug which fits into a modular jack held in place by a wallplate somewhere along the baseboard in one of your rooms (Fig. 5-5).

(A) Insert male plug of base unit in convenient ac outlet.   B) Connect modular jack to telephone receptacle.

Fig. 5-5. Installing cordless phone is easy.

PULSE DIALING

All wireless phones use pulse dialing even though you will find the handset is equipped with a pushbutton panel (pad). This feature makes the wireless phone compatible with both types of telephone lines, those intended for rotary dialing and those designed to accept pushbutton calling. There are some wireless phones, though, made for either pulse or tone dialing.

FEATURES

All wireless telephones are alike in their function, that is to permit use of an in-home telephone at some considerable distance away from it without needing connecting wires, but not all of them have the same features. There is also a considerable variation in price. Usually, the greater the operating range and the more features the higher is the cost. Styling is also a consideration.

Automatic Memory Dialing

Cordless telephones can have the same features as in-home or in-office units. One of these is automatic memory dialing that lets you store a group of 10 or more digits and to dial them with just the touch of a button. Touch another button and it will automatically redial the last number entered.

Range Alarm

It is often difficult to know just how far you can go with the handset portion of a cordless telephone. One method is to run a series of tests to check on the distance. Even though a unit may have an estimated range of 600 feet, whether it will actually have this range depends on some factors outside the control of the cordless telephone manufacturer.

Since home wiring is used by the base station for transmitting signals, the reach of the signal will depend on the type of home wiring installation. If it consists of unshielded Romex cable (and this is usually the case), then the radiated signal should have little difficulty in reaching the handset at its optimum distance. However, if the home wiring is the older style BX cable, with the current-carrying wires enclosed inside the metal shield of the BX, transmitted signal strength will be cut down considerably.

You can overcome the range limitation problem by using an extension cord to connect the base station to its ac outlet. A long cord, having a minimum stretch of 25 feet is preferable. Don't bunch the cord or form it into a tight loop to make it inconspicuous. Instead, mount it in a straight, tight line along the baseboard of a pair of adjoining walls so the extension cord forms a right angle with itself. This will give you signal radiation in two directions.

Range will also be affected by the location of the base station. Keep the base station situated as high as possible, on a second floor in preference to a first, or on a first floor in preference to a basement.

Some cordless phones are equipped with a warning tone that will automatically sound when you reach the limit of their range.

Intercom

If the cordless telephone base station is equipped with an intercom switch you can use it for two-way in-home conversations. If the intercom has full duplex capabilities you will be able to talk back and forth without continually depressing and letting up on the button.

Dialing Flexibility

Some cordless phones are equipped with a switch for selecting pulse signal operation (10 or 20 pulses per second) or pushbutton tone dialing. If the unit also has pure tone activated dialing it will be capable of full central telephone equipment interface.

Memory

You can get a cordless telephone that is memory equipped. Memory capability varies but you can get one that will store up to nine pre-programmed numbers (or more) for automatic dialing.

Multifunction Cordless Telephones

Cordless phones have now been combined with other components, such as the unit illustrated in Fig. 5-6. It has an fm/am radio and a digital clock. The telephone is a full-duplex cordless unit, operating exactly the same as conventional wired phones, so calls can be received and made. The base unit has a charge indicator to show the charge readiness of the battery used in the handset. The cordless handset controls include talk/standby, volume hi/low, and power on/off switches plus a battery-low indicator. The 12-position

Fig. 5-6. Cordless telephone base station equipped with an am/fm radio and a clock. *(Courtesy Uniden Corp. of America)*

keyboard includes a redial function. You can depress the * key for automatic dialing of the last number dialed. There is a mute key for disconnecting the mouthpiece when you want conversation privacy with persons nearby. The clock has a large, 4-digit green LED time display; off/on/auto alarm and time set/lock/alarm switches; sleep, fast/slow and forward/backward switches; display hi-low; and snooze alarm switches.

Volume Control

The base station unit can be speaker equipped with a slide-type control for adjusting the volume.

Dual Range Handset

Generally, when using the handset, the telescoping antenna should be raised. Some, though, are dual-range types having a range of 100 feet indoors without raising the hidden antenna, and five to ten times this range outdoors with the antenna extended.

Decor

The base station, equipped with its small telescoping antenna, may not even look like a telephone, but rather like an attractive box (Fig. 5-7). With the cover of the box opened you can dial outside numbers, page the handset, and communicate without the need for lifting an earpiece. The speaker of the base station is built into the unit so you can have hands-free talks either with the person using the handset or for outside calls. Similarly, you can use the handset to make telephone calls, local or long distance.

Security

Some cordless telephones (Fig. 5-8) are digitally coded, with hundreds of possible codes. These prevent other people from getting their signal into the base of your phone unless they know your code number and have a handset in the same channel. This helps prevent unauthorized use of your base telephone.

The code switch is often an 8-digit type and you will find it (if the cordless system is so equipped) under the rear or the bottom of the base unit and under the keyboard of the handset. The code switch may be covered with a small plastic tab. In one cordless system there are as many as 512 ways in whatever combination you want to match

Fig. 5-7. This base station is housed in an attractive box. *(Courtesy American Telecommunications Corp.)*

your handset and base unit. As an example, if you set your handset to 2 — 3 — 5 of the code switch, you must also set the base unit code switch to the same numbers. Both code switches, base and handset, must not only be turned on but must be set to identical numbers. If you just set one the system will not operate; and, if you set both, but to different numbers, the system will also not operate. While this may be annoying, it is for your own protection against false operation or interference from other cordless telephone systems.

Hold Key Indicator

A hold key indicator avoids the loss of a call when switching between the telephone and intercom functions.

RF Sensitivity Switch

Rf is an abbreviation for radio frequency and is an indication of the channels used by a cordless telephone system. The receivers in that system have a certain amount of sensitivity for the signals they receive. If the sensitivity is too low communications between the base

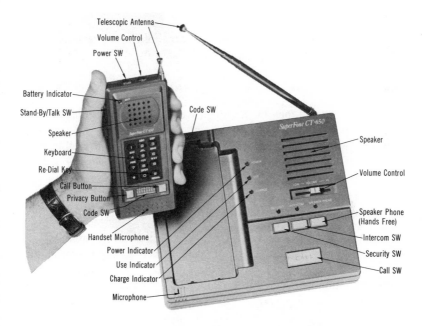

Fig. 5-8. Cordless telephone system uses coding arrangement to prevent unauthorized use. *(Courtesy Samhill Enterprises, Inc.)*

and handset will be poor; if too high, unnecessary electrical interference and electrical noise will be picked up. Some units have a sensitivity switch to control sensitivity. With the switch in its off position, sensitivity is maximum. When the switch is turned on, sensitivity is reduced, but range is reduced as well. There is not much point in having more range than necessary for effective communications.

Buzzer

The remote handset will probably have a buzzer that sounds to indicate a call is coming in or someone at the base is paging the user.

Recorder

You can get a cordless telephone that includes an in-line microcassette recorder that makes possible keeping a permanent record of all incoming and outgoing calls. With the built-in recorder, there is no danger of lost calls or garbled messages. Microcassettes can record up to 30 minutes of conversation on both sides.

Mute

With this control the handset user can hold a conversation with others nearby without having a caller listen in.

Speaker

The base station may be equipped with a speaker to permit hands-free operation, both for incoming calls and also for two-way intercom use. Automatic hands-free vox operation is often desirable when the base is located in a quiet area. Use of a manual override is desirable when the base is in a noisy environment.

Communications Capabilities

The cordless should have a three-way conferencing capability between incoming call, handset and base, as well as any two-way conversation between base and cordless handset; base and incoming calls; incoming calls and cordless handset (or handsets); and cordless handset and called party.

WHERE TO USE A CORDLESS PHONE

You can use a cordless phone in an office, outside a home or inside. Some cordless phones, as indicated earlier, have an intercom feature, so it can be used in place of such a unit.

You can take a phone call wherever you are, provided you are within the transmitting and receiving range of the base telephone. And, since the cordless phone is low-voltage battery operated you can use it safely in the bathroom or in any other situation where water is involved. The cordless telephone remote handset does not have the high electrical hazard of ac operated appliances. This, however, is not true of the base station. That unit *is* ac operated and should be treated with the same care and caution you use on any other component that is connected to the ac line.

A cordless telephone system is ideal for use in the bedroom of someone who is ill or who is not ambulatory. In this way the patient can communicate with all other members of the household by using a remote handset in the bedroom. That handset can not only be used for communications purposes, but for paging and intercom.

PAGING

A cordless system can be used for paging. In the paging mode another person, possibly a member of your family or a friend, can beep you when there's an incoming call for you. You would then receive this alert on the remote. And you can beep the base station just by pushing a call button on the handset. Do not use the cordless phone in connection with a party telephone line.

HANDSET CONTROLS

Not all cordless telephones, base units, and remote handsets have the same number of controls (Fig. 5-9) but those listed here will indicate what you may expect.

Talk Switch

This may be located on the face of the unit near the built-in earpiece. There may also be an illuminated talk indicator.

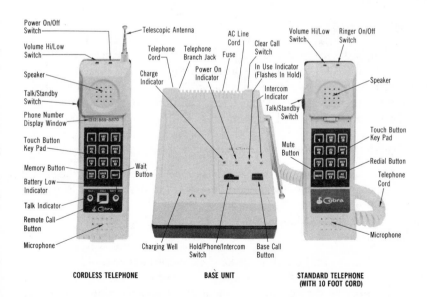

Fig. 5-9. Three-piece cordless telephone system. *(Courtesy Dynascan Corp.)*

Battery-Low Indicator

This indicator glows when the voltage of the battery drops to a level that is too low. In that case it becomes advisable to place the handset into position in the base unit cradle. For many cordless telephones this puts the battery on automatic charge. If the base does not have a charger, an external charger is needed, sold as an optional accessory.

Power Switch

The power on/off switch on the handset is often a slide type. This switch should always be kept in its off position when the handset is not expected to be in use. This will help conserve battery power and make battery recharging needed less often.

Pulse Selector Switch

This control will select either 10 or 20 pulses per second, determined by the characteristics of your telephone line. To decide the setting of the switch call your local telephone utility and ask them. Once you have this information this switch can remain fixed in one position.

Touch Dialing Keys

This is the key pad and will be on the front of the remote handset. It will have keys for all digits from 0 through 9, or a total of 10 keys. There will also be two additional keys, both at the bottom of the pad, with one at the extreme left, the other at the extreme right. These special keys will be marked # and *.

On-Hook Function Key

Some cordless units have a provision for mounting on a hook. By depressing the On-hook function key you can dial without lifting the remote unit.

Redial Key

By depressing the rightmost key (mentioned above under the heading of Touch Dialing Keys) on the bottom row of the telephone pad you can have the remote unit memorize one telephone number for subsequent automatic dialing when this key is depressed.

Microphone

The microphone in the handset is usually an electret condenser type and is built into the component.

CONTROLS ON THE BASE UNIT

Like the handset, the base unit is also equipped with a number of controls.

Power Indicator Lights

These will glow when the ac line cord for the base unit is plugged into an active outlet. If the indicator does not light, try a different outlet.

Recharging Indicator

This indicator glows when the remote handset is positioned in the cradle. It indicates that the nicad batteries in the handset are being recharged.

Recharging Terminals

These terminals make contact with a similar pair of terminals on the handset remote unit. When this contact is made the battery recharging process begins.

Call Button

Depress this pushbutton when you are at the base unit and want to call the remote handset.

In-Use Indicator

This indicator glows when the unit is in its ready-to-talk mode.

Volume Control

A high/low volume control lets you turn the sound up or down. This feature is ideal when talking from a noisy location. The volume control may be only on the base unit, on the handset, or on both.

Secure Override

Some wireless systems are equipped with this feature, an advantage when using two handsets. The secure override automatically releases the telephone for the user of a second handset, even when the primary handset is hung up in the base station cradle.

Extra Handsets

It is possible to use more than one handset, just as it is possible to have extension telephones in the home. While this offers the convenience of having handsets in various locations it does not mean the base station can be used as a switchboard. The base station can handle signals from one handset at a time. Using a pair of handsets can cause confusion and interference unless there is some way you can use the handsets in a simplex mode, that is, operating the handsets one at a time, not simultaneously.

CORDLESS-TELEPHONE RANGE EXTENDER

The usual cordless telephone has a range that is more than adequate for average use and for most applications the range is generally greater than needed. However, it is possible to extend the transmission and receiving distance by equipping the cordless telephone system with supplementary antennas as illustrated in Fig. 5-10. The antenna used for receiving signals consists of a vertical dipole tuned to 49 MHz. It should be mounted as high and free of the ground as possible, possibly by installing it on the roof of the house. It can be attached to the side of the house or fastened to a chimney. The transmitting antenna is a long wire. The antennas are connected to an adapter and from the adapter a wire is clipped to the existing antenna on the cordless phone.

The connections between the antennas are via coaxial cable. No internal connections to the cordless telephone are required and the use of the antenna system will not void the cordless telephone warranty.

The range extender is used in connection with the base station. When installing the antennas be sure to stay clear of any power lines for touching these can result in electrocution.

The unit shown in Fig. 5-11 is just one of a number of cordless phone range extenders designed for use on all 1.7/49 MHz phones.

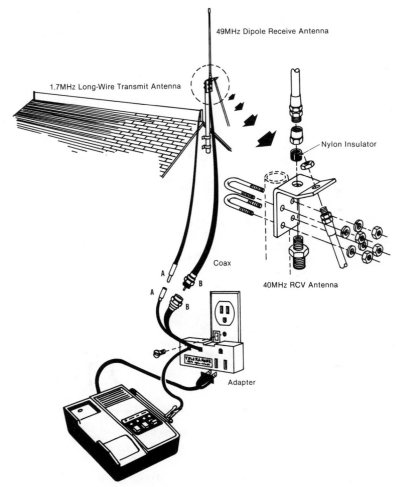

Fig. 5-10. Antenna installation for extending the range of a cordless phone. *(Courtesy Valor Enterprises, Inc.)*

Still another range extender antenna is shown in the photo in Fig. 5-12. For strongest signal radiation the antenna should be mounted above or a minimum of 18 feet away from any metal object. The antenna can be mounted indoors or out, but for best performance put the antenna as high as possible. It can be located as much as 50 feet from the base unit.

Note that antenna range extenders are designed to be used with the

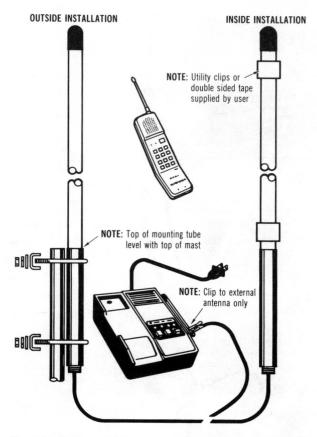

Fig. 5-11. Range extender antenna for cordless telephones.
(Courtesy Shakespeare Co.)

base station and not with the handset. The base station is a fixed position device, but the remote handset is not.

CORDLESS TELEPHONE PROBLEMS

Cordless telephones, particularly those that are extremely sensitive, that is units that have a very long receiving and transmitting range, possibly 700 feet or more, can sometimes be triggered into operation by a nearby cordless unit.

Cordless telephone systems can pick up signal interference in two

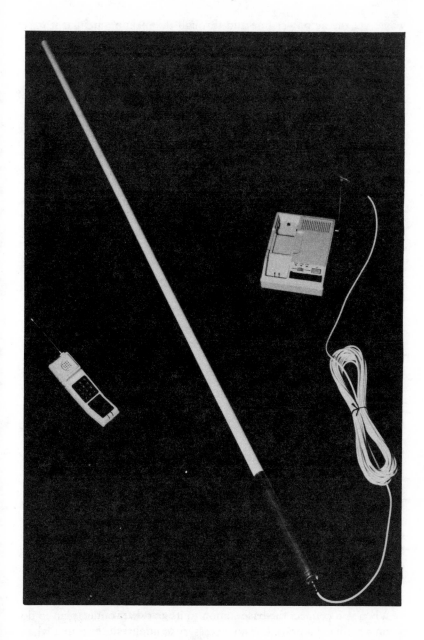

Fig. 5-12. Antenna for **extending** cordless telephone range can be used indoors or out. *(Courtesy Shakespeare Co.)*

ways: via the ac power line and through the pickup antenna. It may take a little detective work to determine the cause. Thus, electrical interference pickup may be due to fluorescent fixtures, electric drills, a hairdryer, etc. If any one or all of these are operating, turn them off one at a time to determine if the interference has disappeared.

Theoretically fm receivers systems should be free of electrical interference, but this depends on the amount and kind of noise reduction circuitry used in the components.

This telephone system, either the remote or base station, can also behave in an erratic fashion, produced by electrical noises, possibly generated by a passing truck or car. Since the base unit is connected to the ac power line, electrical noise impulses can travel from that line into the base station. The noise signals can not only affect the base, but the remote handset as well, since the base transmitter is capable of radiating the noise impulses.

Noise filters are available that can solve this problem. The filter is inserted between the base unit and the ac power line. The base power line cord is put into an outlet mounted on the filter while a connecting cord from the filter is plugged into the power line.

Sometimes lightning striking a telephone line will produce a surge of current that will pass from the line into the base unit. There it can possibly damage some of the circuits, possibly not enough to cause them to stop working, but to function erratically.

For best communications results between the base station and the remote handset, do not install the base unit near any large metal masses, such as a refrigerator, all metal kitchen cabinets, a furnace, etc. These large metal objects are signal absorptive, and so they "soak up" signals being received by the base station. The same effect will be noticed in rooms that use metallic wallpaper covering or which use aluminum backed insulation in the walls. A good method would be to install the base station in the area most convenient for you and then to run a test. In performing this test the remote handset should be tried in all of the possible locations you plan to use for it. If results are not satisfactory try moving the base station to some other area. The antenna on the base station and also on the remote handset should always be fully extended. The antennas can be recessed, if you wish, when you do not plan to use the system.

When you connect the base station to its ac power outlet, plan to do so on a permanent basis. It is not necessary to unplug the base unit when you are not using the cordless telephone. The base unit is designed for continuous operation and uses very little electrical energy.

Your base unit may have a line voltage selector reading 110V – 120V and 220V – 240V. For homes in the U.S. the line voltage is 110V – 120V so be sure to set the line voltage selector to these numbers. 110V – 120V means your line voltage is somewhere between 110 and 120 volts ac.

APPLICATIONS OF THE CORDLESS TELEPHONE

Just like regular telephones, the cordless type can be used either in the home or in the office.

Personal Use

The cordless phone, even those with a minimum range of possibly 300 feet, can be used anywhere in the home (including the garage) as a portable extension phone and you can use the remote handset to make telephone calls anywhere you wish, local or long distance. You can also receive calls, so you do not miss any if you are not within reach or hearing distance of your main, base station telephone.

A cordless telephone can be used by recuperating patients and invalids as a bedside communications system. The cordless eliminates the need for installing a special telephone in the sick bedroom. An extra telephone, if owned and installed by your local telephone utility, means an increase in your monthly telephone bill. The cordless remote eliminates these charges. A cordless telephone also means you need not install any telephone wiring in the invalid's bedroom. Further, if the invalid is ambulatory, the cordless can be carried anywhere the invalid wishes to go, including the bathroom.

With a cordless telephone you need not be concerned you may miss important calls received inside the house when you are outside, in the yard, patio, poolside, garage, or on your front walk. The cordless will alert you as soon as a call is received. Further, you can answer that call without making a frantic dash into the house. And you can make telephone calls from wherever you are.

If you plan to go on a trip and leave your house unoccupied you can deposit your cordless telephone with a friendly, nearby neighbor who will be willing to answer your telephone calls and take messages if necessary. It is advisable to check to make sure this neighbor is within the operating range of your cordless telephone system. An alternative, of course, is to install a telephone answering machine, described later in Chapter 7.

Commercial Use

Restaurants and clubs can provide their customers with convenient access to telephone lines. A patron, receiving a telephone call, need not leave his table and can not only answer a call directly, but can also make outgoing calls. Restaurant and club owners need not install telephone lines leading to tables.

A cordless system can be used for interoffice or factory communications. Any production person on the floor, or a foreman or supervisor can easily contact a manager and discuss production problems. The use of the cordless telephone here saves considerable unnecessary running around. The telephone is faster than physical movement.

If a handset telephone is available at a construction site wasted time can be eliminated by communications between supervisors and workmen. In hospitals, rest homes and homes for the aged, the cordless offers a quick means of communications between nurses, doctors, staff members, and office personnel.

CORDLESS TECHNOLOGY

A cordless telephone setup can be as electronically sophisticated as any other kind of telephone system. The unit in Fig. 5-13 is two complete telephones in one system. It has a switchable universal rotary and DTMF (tone) arrangement that provides the user with a 32-number memory dialer with automatic last number redial. Its operating range is rated at 1000 feet depending on local conditions.

Other features include microprocessor circuitry, and full duplex intercom between the base and remote, enabling either unit to activate the intercom mode while putting an outside call on hold. The system automatically releases the hold.

The base unit will accept up to five separate remote units and the remotes can be paged from the base unit either individually or all at once.

The system's 32-number memory dialer and last number redial can be accessed from both the base unit and the remote. The user can switch automatically between DTMF tone and rotary pulse dialing. It also allows switching from pulse to DTMF for MCI™ and Sprint™ type calls. In addition, over 1200 privacy/security codes per channel permit the programming of a personal code system.

There is also a microcomputer voice synthesizer. This aids in pro-

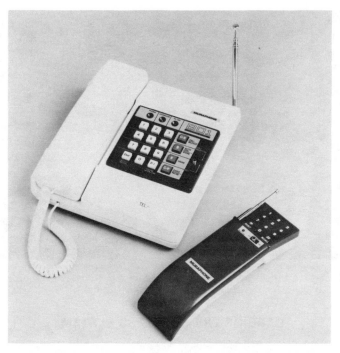

Fig. 5-13. Cordless telephone has voice synthesizer that aids in programming and reads numbers stored in memory. *(Courtesy Mura Corp.)*

gramming and will read out all numbers stored in the memory, either individually or all of them sequentially. This voice synthesized readout is available at both the base and remote units. The system is also compatible with PABX office systems and can access the entire range of information storage and retrieval through its DTMF tone dialing system.

CORDLESS WITH RECORDER

A continuing process of product integration is taking place in telecommunications. Thus, the unit illustrated in Fig. 5-14 is a cordless telephone wth a 700-foot range which features a builtin microcassette recorder for monitoring incoming and outgoing calls. The photo shows the way in which the remote is put into its base cradle for recharging.

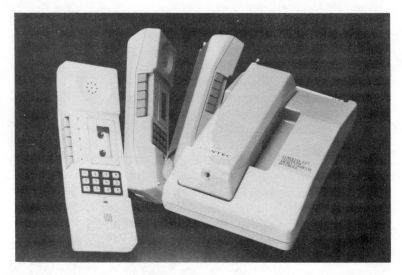

Fig. 5-14. Cordless telephone equipped with a microcassette recorder.
(Courtesy Contec Electronics Co.)

CARE OF THE CORDLESS SYSTEM

While the remote handset is designed to be used either in or out-doors, it should be kept in your house overnight to protect it against the weather. As indicated earlier, it is a good idea to put the handset into its charging cradle and keep it there overnight so as to maintain the batteries in a fully charged condition.

If you plan not to use the cordless system for a period of time, possibly weeks or more, remove the batteries from the handset and store them separately. Disconnect the base unit from the ac power line. It is not necessary to remove the base station telephone from the phone outlet.

MALFUNCTIONING

If, after you first set up your cordless system it does not work, put the batteries on charge overnight and then try again. If, however, after following the manufacturer's instruction sheet (packed in with the product) you cannot get the cordless unit to work, disconnect it from the telephone line until repairs have been completed.

Your cordless telephone should have no adverse effect on the telephone lines. However, if for any reason the cordless base is causing telephone line trouble your local telephone utility may disconnect your phone service temporarily. They cannot do so without giving you advance notice, unless the situation is an emergency.

If the telephone line problem is indeed due to faulty operation of your cordless telephone system they must notify you and give you an opportunity to correct the problem. They must also advise you of your right to file a complaint with the FCC against them. But before you do so make sure of your facts and also be sure the manufacturer of your cordless telephone has had an opportunity to examine the equipment.

Always keep your sales slip, not only as proof of purchase but as proof of the *date* of purchase. This will enable you to establish your warranty rights. You should follow this procedure, not only with cordless telephones, but with any other type of telephones or phone systems, or with add-on units.

CHAPTER 6

GUIDE TO OWNING YOUR OWN TELEPHONE

Because telephones have customarily been rented, not sold, private ownership has raised many questions. However, the right to own your own telephone carries with it some responsibilities and some necessary precautions.

Is it necessary for me to buy at least one telephone from the telephone company? No. All of the telephones you have in your home, including any add-on accessory units, can be components you have purchased elsewhere. However, any equipment that interfaces with telephone lines must be FCC approved and must be reported to your local phone company.

Why is it necessary for me to pay my local phone company anything if I own my own telephone? You pay them for the use of their lines, for supplying power to your telephones, for connecting you, using their equipment, via telephone, with anyone with whom you want to make contact. Telephone companies must maintain their lines, repair them, must supplement them by adding new lines when necessary, must invest in expensive switching equipment, must supply long distance service, and must research new possibilities in tele-communications.

If I have a choice of tone dialing vs rotary dialing, which should I select? Tone dialing has several advantages over rotary dialing. It is faster. There are also a number of different kinds of services which can only be had when using a tone telephone. These include computerized banking and shopping, access to long distance lines other than those supplied by your local utility (lines such as Sprint™, MCI™, etc.). It is also possible rotary dialing may later be eliminated.

Should I keep the sales slip when I buy a new telephone?
Definitely. This establishes the beginning date of your warranty. You
will need this slip in the event your telephone becomes defective
within the time period of the warranty. You may also need the sales
slip as a business expense deduction when computing your taxes.

Are there any advantages in buying used telephones? The chief,
and possibly the only, advantage is the saving you may make over the
cost of a new telephone. Used telephones aren't often supplied with a
warranty. You may also have a problem in getting the telephone repaired if you don't know the name of the manufacturer. Not all used
telephones are FCC approved, and in that case you may not connect
such telephones to your line. Older telephones may use standard instead of modular plugs and so this may mean you will need to buy an
adapter. Finally, newer telephones are gradually being equipped with
more and more features, something you may not find in a used telephone. The saving you make in buying a used telephone may not be
enough to justify taking all of these risks.

Will I have any trouble connecting my new telephone? Not if it is
an FCC approved type. FCC approved plugs and jacks are required
so users can avoid connection difficulties.

Will my telephone company install phone jacks for me? Yes, and
they will install them where you want them. There is a charge for this
service. It is more economical to have all the jacks you think you may
need installed at one time. If the installer must make a number of
trips if you should happen to decide later you want more jacks, you
will need to pay an additional charge.

Can I connect my telephone to a party line? Not if you own your
own telephone. According to FCC rules, customer-owned telephones
can be connected only to private lines, not to party lines or coin lines.

Are modular plugs and jacks standardized? Yes, these are standardized throughout the U.S. and they are fully compatible. Compatibility
means that any FCC approved telephone you buy will have a modular plug that will fit into any modular jack. The jack may be contained in a variety of different housings, but this will not affect its
ability to mate with a modular plug.

If I move can I take my telephone with me? Yes, if you own the
telephone. If it is a telephone utility telephone — that is, one that is
rented from that utility — you should notify them and make arrangements for its return. It is quite possible you have a refundable
deposit on that telephone.

If I move and take my phone with me, will it fit into jacks in my

new home? If you have an FCC approved phone equipped with a modular plug it will fit into a modular jack anywhere. This assumes your new home has modular jacks.

Can I connect a phone with a modular jack to an older 4-prong style jack? Yes. Adapters are available to convert modular jacks so they will plug into standard jacks.

Is it legal to own my own telephone (or telephones)? Owning your own telephone has been legal since October 1977. At that time the FCC (Federal Communications Commission) made it permissible for everyone to connect their own telephones to telephone company lines. Prior to this telephone companies insisted on the use of a protective connecting arrangement. The original FCC order was issued in November 1975 but this order was challenged in the courts. FCC regulations regarding phone ownership went into effect almost two years later.

What is a registered telephone? All telephones, and that includes those made by telephone companies as well as competing companies, must not only be registered with the FCC but must be labeled with an FCC registration number and also a ringer equivalence number. These numbers indicate that the phone complies with FCC requirements.

Can I buy and use a phone made in a foreign country? Yes, provided the phone has an FCC registration number and a ringer equivalence number.

Is it difficult to install your own telephone? It is as easy to install a telephone as it is to plug a lamp into an outlet. All new phones that carry an FCC registration number have a standard or modular plug that fits in or mates with a standard or modular telephone jack. It is possible you may need to use an adapter.

Must I inform my local telephone company that I am putting in my own phone? Yes. This is a requirement of the FCC and you must not only notify them but you must supply them with two bits of information — the registration number and the ringer equivalence number. A phone that does not have this data is not suitable.

Must the manufacturer of a telephone supply me with the registration number and the ringer equivalence number of the telephone I buy? Yes. This information will generally be printed on the shipping container or printed on a paper packed in with the telephone or it may be on the telephone. If this information is not supplied, do not buy the unit.

Can I install my own telephone jack? Whether you are legally allowed to do so depends on the state in which you live. Some states permit it, others do not. Installing a jack means you are also going to put in your own wiring. Telephone your local state public utilities commission to learn of the rules that apply in your community.

Am I required to use telephones previously installed by the telephone company? No. You can replace any phone or phones, provided they are FCC registered and that you notify your telephone company.

How many phones can I install? The usual maximum is 4 or 5 phones, assuming all these are ringer equipped. It takes electrical current to ring phones and the amount of current being supplied may not be enough. It is advisable to check with your phone company to learn the practical limit. Not all phones use ringers. You could have one phone with a ringer and several phones without this device. You can exceed the telephone limit by using phones without ringers.

Is telephone registration a guarantee of quality? No. The purpose of FCC registration is to make certain that the telephone(s) to be connected will not cause damage to the telephone system. It is always best to buy a new phone from an established company with a reputation to maintain. Always be sure to buy a telephone that is supplied with a warranty. After you install the telephone check the phone by calling in from the outside. In that way you will get some idea of what your telephone will sound like to others. Before buying a telephone try to get some information not only about speech fidelity, but about the durability of the phone and its life expectancy. Most telephones being sold today are backed by a warranty. A quality phone will have a one year limited warranty.

Is it economically worthwhile to own a telephone instead of renting from a phone company? You can save money if the phone you buy is the same size, style and type as the telephone you are replacing. The reason for this is that telephone prices can range from as little as $20 to as much as $250, and possibly more. The more you spend for a phone the longer it will take to make buying your own phone economically feasible. How much you will save monthly depends on the monthly rental charge that was made by your local utility. You can expect to save approximately $2 a month. If you buy a phone for $200 it would take about 8 years. If you buy a phone for $20 it would take only 10 months.

Must I replace my telephone with an identical style? No. As long as the telephone meets the FCC's requirements, you can use any style of telephone you wish.

Must I replace my telephone with one having the same features?
No. One of the advantages of shopping for your own phone is that
you can buy one having features that may not be available with telephone company phones.

What are some of the features that are available? Telephone features are described in Chapter 3.

Must I use my local telephone company to make local calls? Yes.

**Must I use my local telephone company to make long distance
calls?** No. While your local telephone company has a monopoly on
local calls, they do have competition for long distance calling. Under
certain circumstances, calls can be made through competing companies.

**Can I get a discount or rebate on my telephone bills if I own tele-
phone company stock?** No.

**If I buy my own telephone(s) will my telephone company take the
responsibility for repairing them?** No.

**In wiring my home for additional telephones, can I use any kind of
wire I wish?** No. There are certain standards you must follow.

**Where can I get more information about telephone wiring stan-
dards?** Your local telephone company has available literature on this
subject. Ask to see *Standards for Customer-Owned Telephone Premises Wiring.*

**If my home is already wired, can I replace it or buy it to avoid a
monthly wiring charge?** You can replace the wiring, provided you follow wiring standards. Whether you can buy existing wire depends on
your own local telephone company. Some permit it; others do not.

Where can I buy wire and telephone accessories? These are available at dedicated telephone stores and stores that sell electronic parts.

What is a dedicated telephone store? This is a store that sells telephone and telephone accessories only.

**My new house contains wire and phone jacks that have been dis-
connected by the phone company. Can I use them?** Yes, if the wire
and jacks were installed by the previous owner. Yes, if the wire and
jacks were installed by the telephone company and have been aban-
doned by them.

**Am I required to hire a wiring expert or someone from the tele-
phone company to install my telephone wiring?** No. You can do the
wiring installation yourself, or you can hire someone of your own
choice. You can also take advantage of information on how to install
premises wire available at the offices of your local public telephone
company.

What is the advantage of installing my own premises wiring? This will reduce your telephone bill since you will no longer be charged for the use of such wiring.

Why must I notify the telephone company that I intend to put in my own premises wiring? There are two reasons for doing so. You want the rental charge removed from your telephone bill. The phone company must also know about the wiring in the event of reported trouble on your phone line.

What is a demarcation jack? This is a jack located inside your premises to which the incoming telephone wires are connected. It can be considered as any point to which a telephone can be attached.

What is the significance of a demarcation jack? It is the starting point for the connection of self-installed premises wire.

I don't know which jack to select as my demarcation jack. Can I ask the telephone company to install one for me for that purpose? Yes, but you would be required to pay a one-time charge for this service.

My home is equipped with all the on-premises wiring I need and I would rather not get involved with wiring installation. But I also do not want to pay rental charges. Can I buy my wiring from my local telephone company? The policy of selling premises wiring is a policy established by each individual telephone company. The only thing you can do is to discuss this with your local phone company.

Can anyone own their own telephones and on-premises wiring? Yes, provided these are for nonparty line service.

Suppose I install my own-premises wiring, can I still use telephones supplied by my phone company? Yes, but your savings would not be as large.

I do not own my own home but live in an apartment. Can I buy and install my own telephones? Yes, and these remain your property when you move, in the same sense as any of your other appliances.

I rent an apartment. Can I install on-premises wiring? Quite possibly under the terms of your lease any changes or improvements you make in or on the apartment become the property of your landlord when you move. In any event your lease probably prohibits you from doing so, or requires you consult with your landlord first.

I am on a party line. Why am I not permitted to own my own telephones and on-premises wiring? Party lines require special wiring, ringing and toll charging methods.

**I am on a party line and am thinking of switching to private line service. This will cost me more but I thought I could save by installing my own premises wiring and buying my own telephones. How much could I save? It is not possible to answer this question since it all depends on the number and kind of telephones you plan to have. However, you can do the necessary arithmetic easily enough. You know how much you presently pay for party line service. Find out from your telephone company what they will charge for private line service, assuming you do your own on-premises wiring and buy your own telephones. You can then make a comparison between the cost for party line service vs private line service.

What is the target date for the deregulation of the Bell Telephone System? Under the settlement reached by the Justice Department and AT&T, the divestiture must be completed by Feb. 24, 1984.

Can I buy telephones from the Bell Telephone Company? Yes, their phones are available through PhoneCenter stores and through various other outlets, such as Sears.

Must I notify my telephone company if I buy a cordless phone? Since a cordless telephone includes a base or fixed position telephone you must notify the phone company about the in-home phone. However, this does not apply to the mobile portion of the cordless system.

What is meant by a cordless telephone channel? This is the operating frequency of the cordless phone.

How many channels are available for cordless telephones? Five.

Is there a possibility that this number of channels will be increased? The FCC has been asked to increase the number of available channels for wireless phone use from five to twenty-five.

After I buy and install new phones, what can I do with my old telephones? Since your old phones most probably are those you rented from your local telephone company, they must be returned. You can have them picked up by the utility, but you will pay a service charge for doing so. In some states the utility will pay you for each phone returned.

What is the fastest way of getting emergency call numbers, such as fire, hospital and police? Some utilities supply these, printed on a self-stick slip of paper which can easily attach to the front or side of the telephone base. If not, you can get the numbers from your directory. Type or write them neatly and put them on the phone or on some surface near the phone. If you cover the emergency phone listing with transparent plastic tape you will keep the numbers from becoming accidentally smeared or dirty.

What is a Universal Dialing Telephone? This is a pushbutton telephone that can work on both types of telephone lines — lines intended for rotary dial operation or lines intended for pushbutton use. The universal dialing telephone is also known as the universal dialing keyset (UDK) and is also called "pulse dialing." A UDK telephone looks similar to a tone signaling phone since its pad is a pushbutton type.

Although this telephone does have pushbuttons, the electronics inside the telephone takes the pushbutton entry and converts it to a rotary dial type entry. This is referred to as *dial pulsing.*

As far as the telephone company is concerned is there a difference between dial pulsing and rotary dialing? No. Dial pulsing signals the telephone company's Central Office just as a rotary dial does.

Are there any advantages in using a telephone having the UDK feature? Yes — there are two important advantages. Telephone companies charge more for pushbutton service, whether they supply the telephones or not. With a UDK you can have the advantage of a pushbutton telephone without paying an extra service charge. Not all telephone companies supply a touch calling service. This means their subscribers must use rotary dial telephones. However, you can avoid this by using a UDK type telephone.

What is the best telephone to buy? There is no such thing as best, just as there is no best kind of car to buy. The kind of telephone you should buy depends on your budget, the kind of phone you prefer, and the number of telephone features, with the price going up for an increased number. What you will pay for a telephone also depends on how good a comparison shopper you are. Because telephones are now being sold by a number of independent manufacturers, there is greater price competition, so it is sometimes worthwhile waiting for a special sale.

Do I need to use a soldering iron or other tool for connecting my phones to telephone company lines? No. There is no need to do any soldering. There are only two types of phone outlets — standard and modular. In either case, no matter what type of terminals your phones have, you can connect them, with the help of an adapter, if necessary, but without the use of tools of any kind and that includes a soldering iron.

Should telephones ever be sanitized? If you are a heavy smoker and use the telephone all the time it is possible for the mouthpiece to acquire an odor. Washing the mouthpiece from time to time with a mild detergent should remove it. Don't bother with antiseptics.

Am I permitted to install a telephone amplifier? You can install any equipment you wish that doesn't interface with the telephone lines.

How can I calculate how much I would save by installing my own telephones and on-premises wiring? It isn't possible to make an exact calculation but you can get a reasonably good approximation. The savings you will make depend on the applicable rates at the time you make the change and on the type of service to which you are a subscriber. The chart shown in Chart 6-1 shows how much you can save (approximately) with your own on-premises wiring and your own telephones.

Chart 6-1. Approximate Savings with Your Own On-Premises Wiring and Telephones. Savings Will Vary From One Telephone Utility to Another. Applicable Taxes Are Extra.

Save with Your Own Wire				
No. of Phones Connected	1	2	3	4
Your savings are:				
Your own wire -Monthly -Annual	-0-* -0-*	1.70 20.40	3.40 40.80	5.10 61.20
*You must have one link of company-provided wire and pay for it monthly.				
Save with Your Own Phones				
(Without "Touch-tone":) Your own "standard" phone -Monthly -Annual	1.85 22.20	3.70 44.40	5.55 66.60	7.40 88.80
Your own "Princess" type phone -Monthly -Annual	4.91 58.92	9.82 117.84	14.73 176.76	19.64 235.68
Your own "Trimline" type phone -Monthly -Annual	5.92 71.04	11.84 142.08	17.76 213.12	23.68 284.16
With "Touch-tone": Additional savings are: -Monthly -Annual	1.06 12.72	2.12 25.44	3.18 38.16	4.24 50.88

Sometimes it is necessary for me to keep the telephone from ringing. Is there any harm in disconnecting it? Modular telephones are
very easy to connect and to disconnect, so easy in fact that some
telephone users remove the connecting plug from its wallplate so as
to obtain telephone privacy. The problem with using this method is
that the connecting module is not made for such work and will even-
tually break. Some manufacturers do suggest this procedure but it is
not recommended. A better method is to use a telephone that has a
ringer disconnect switch. This is easier, quicker, and better. It is certainly much more convenient. You can also get an accessory that will
give you the equivalent of a disconnect switch.

What is a PC board? This is a printed circuit board. It is usually a
fairly small (measured in inches) board on which conductors are
etched. The electrical parts are mounted on the board with the inter-
connections made by the etched wiring. Printed circuit boards have
replaced point-to-point wiring, a method in which components were
mounted on a base or chassis and then connected by wires. The advantage of a PC board is that wiring errors are eliminated, less space
is required, overall weight is decreased, and manufacturing costs are
lowered.

What is a chip? A chip is a tiny bit of silicon on which semiconductors such as diodes and transistors can be etched. Thousands of these
can be put on a chip about the size of a fingernail, or less.

What is an integrated circuit? An integrated circuit is a complete
circuit, with all its parts, designed to perform specific functions. Integrated circuits are often made on chips.

What is a microchip? A microchip is an extremely small integrated
circuit. One example is the *Belmac-32A* (Bellmac is a trademark of
Western Electric). This is a new microprocessor developed by Bell
Laboratories and is packed onto a silicon chip about the size of a
dime. It contains almost 150,000 transistors and can process 32 bits
of information at once, easily the equivalent of a computer that
would have occupied a large room just a decade ago.

Bell Labs designed the Bellmac-32A specifically for telecommunications uses. Microprocessors like this one could serve as controllers
for information subsystems — for energy, security, fire and kitchen
appliance control — as well as for home entertainment and home or
business information centers.

Is it possible to control home appliances by telephone? Yes. By
using your telephone you can control house systems and appliances
by remote control. The system shown in Fig. 6-1 allows com-

Fig. 6-1. Kitchen appliances and house systems can be controlled by phone.
(Courtesy General Electric Co.)

puterized, preprogrammed, at home or phone remote control and monitoring of heating and cooling systems, kitchen appliances, lighting, security and smoke alarms, phone message devices, and home entertainment systems.

The systems module consisting of a video terminal (which can also be used as a regular TV receiver) and keyboard, can be built-in, or can be a free-standing device.

CHAPTER 7

TELEPHONE ANSWERING MACHINES

A telephone answering machine is exactly what its name suggests — a device for answering your telephone in your absence. But this description is much too limited for you can use it to screen your calls or work as a dictating machine. And you can "beep" it or voice-actuate it using any telephone anywhere in the world, ordering it to give you your messages, to erase them, or to keep a record of a selected few. Like other electronic components, what you will pay for an answering machine depends on its varsatility, that is, on the number of features it has. But, regardless of its price there is one thing you can be sure of, it will remain on duty 24 hours a day, 7 days a week, subject only to any possible failure of the ac power lines.

BUILT-IN TELEPHONE

Some telephone answering machines are integrated with a telephone (Fig. 7-1); others are not (Fig. 7-2). Having a unit with a built-in telephone is a convenience since it becomes easy to control the recording of all your outgoing messages as well as incoming calls. The best arrangement is a telephone with a universal dialing keyset. In this way you can connect to any phone line, whether that line is intended only for pulse dialing or rotary dialing.

Another advantage of the built-in telephone is it gives you the convenience of an extension phone. Also, since you now have an extra telephone you can apply for two listings, a convenience in the event you keep your phone lines fairly busy. This is an additional telephone expense, but it may be necessary for your particular activities.

Compatibility

Prior to buying a telephone answering machine make sure the device is compatible with the telephone you plan to use. Compatibility is assured if the answering machine is telephone equipped. And if the answering machine is telephone equipped it is advantageous to have the telephone capable of pushbutton (tone) dialing or pulse (rotary) dialing, generally handled by a switch on the unit. If not, check with your local phone utility to learn which type of dialing is acceptable.

Battery Backup

If the answering machine is telephone equipped then that phone can have any or all of the telephone features described previously.

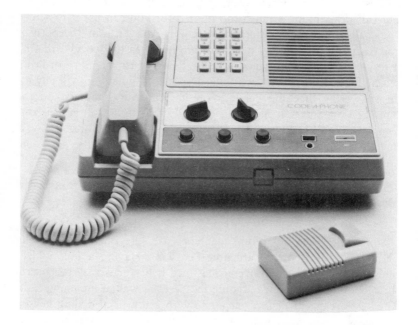

Fig. 7-1. Integrated telephone and telephone answering machine. It can accept up to 20 messages of 40 seconds each. You can use a "beeper" type coder to cancel messages, record memos, or repeat messages. An LED light flashes when messages have been received and glows when the unit is on and ready to receive calls. *(Courtesy Code-A Phone, Ford Industries)*

Fig. 7-2. This TAD has a beeperless remote function. The user simply commands the unit by voice, using any type of telephone, to playback messages, rewind, or to change announcements. The microprocessor controlled, dual mini cassette unit also features message scan, repeat, two-way recording and a jack to accommodate a cordless telephone for call screening. *(Courtesy Pathcom, Inc.)*

Some of these machines are supplied with a battery backup, advantageous if the setup is moved to a new location. With a battery backup, numbers stored in the memory are retained.

ABBREVIATIONS AND ACRONYMS

No sooner does a new technology develop but it is accompanied by a number of abbreviations or acronyms. These soon become so commonly used the assumption is made that everyone knows what they mean. You will, for example, find these abbreviations used in the instruction manual accompanying the answering machine you buy, often without an explanation of what they mean. A few are indicated here; others are in the glossary following the last chapter.

OGM
  Outgoing message tape.
ICM
  Incoming message tape.

VOX
  Voice actuated.
TAD
  Telephone Answering Device.
MIC
  Microphone. This is sometimes abbreviated as mike.

TAD

A telephone answering machine is sometimes referred to as a *TAD*, an acronym for *T*elephone *A*nswering *D*evice. There are telephone answering services available that perform the functions of a TAD but these are relatively expensive. A TAD can do the same work and involves just a one-time payment in about the $100 to less than $400 range.

A TAD is actually a phone-operated tape recorder and so many of the functions of a tape recorder/player also appear in connection with a TAD, such as playback, record, erase, forward and reverse. And, like any tape recorder/player, a TAD will accept an audio signal from the telephone lines and will record it and play it back on demand. Moving in the other direction, the TAD supplies a tape recorded audio signal to the phone lines.

TELEPHONE RECORDING

If all you want to do is to record your telephone conversations you do not even need a TAD if you already have a cassette tape recorder. A unit such as the one shown in Fig. 7-3 will let you tape all phone calls, both incoming and outgoing. It does not require any source of power, either batteries or house current, and works undetected and automatically. It works by starting and stopping your tape recorder every time your telephone or extension is used. However, Federal law requires that the calling party to the conversation be informed the call is being monitored. (A comparable unit was shown earlier in Chapter 4.)

In telephony, the word *monitor* has several different meanings. Thus, monitoring can mean listening, and so you can monitor a conversation just by hearing it. Monitoring can also mean recording. You can monitor a telephone conversation by recording it without

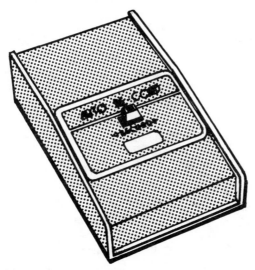

Fig. 7-3. Automatic telephone recording unit tapes all phone calls on your own cassette tape recorder. *(Courtesy Telco Products Corp.)*

necessarily listening to it at the same time. The assumption is made that if you monitor a telephone call, you will ultimately listen to it.

CALL MONITORING

One of the advantages of a TAD is that you need not answer a telephone call if you do not want to. The TAD can answer for you while, at the same time, you monitor that call. At the same time you can override the monitoring function if, after listening to the message for a few seconds, you decide to take over. If you decide not to override the machine will still take a message for you.

PHONE SONGS

One of the problems with a TAD is the caller may not leave a message and in this way eliminates the usefulness of the answering machine as far as that particular call is concerned. There are various reasons why a caller will not reply to an answering machine. Some people become resentful about talking to a TAD. They may feel slighted, hurt, or annoyed. Perhaps it is "microphone fright" but whatever the psychological reason may be, the result is always the same — a hang-up.

To overcome this block some individuals may have about talking to an answering machine, you could try using a Phone Song. These are cassette recordings of originally composed musical greetings designed to be replayed on any TAD. There are various selections you can make. Some are business or office oriented, others are more appropriate for personal or home use. The purpose of a Phone Song is not to entertain. It is a deliberate, psychologically subtle means of manipulating callers to do what answering machine owners want most, and that is to leave a message. Market statistics indicate the use of a Phone Song results in a 15-percent increase in mess ages left, based on an average of 330 calls per year among those surveyed.

Another solution to the problem of telephone hang-ups is to use a female voice on the announcement tape. Make sure the voice is soft, not strident. It should be sufficiently loud so the caller does not need to strain to hear. The voice should articulate well, so every word of the message can be clearly understood, since it must be understood with the first playing. The message itself should be concise. Prior to recording the announcement message, write it, and then edit it carefully, removing any unnecessary words. It is essential to remember the caller cannot interrupt the announcement.

If the announcement must be a long one, possibly using the entire side of the announcement tape, it would be advisable to alert the caller to that fact. Thus, you may not want to receive a message for recording and just want the caller to hear what you have to say.

It is also helpful to remember the caller is paying for the call, and if it is long distance, might not appreciate listening to a lengthy announcement. The caller can always hang up but this would defeat the purpose of the TAD.

MESSAGE RECORDING TIME

TADs can supply various amounts of recording time. Thus, a machine might have the capability of recording up to 20 thirty-second messages or 50 thirty five-second messages or 20 forty-second messages. Some can record messages up to 30 minutes long or you can select a 45-second maximum. The total amount of recording time depends on the tape that is used.

TAPES FOR TADS

Two types of tapes are used for TADs. One of these is the standard cassette and is the same kind of cassette used for recording and playback audio cassette decks or portable cassette players. The total recording and playback time is indicated on the shell of the cassette and is a C number. Thus, a C-60 has a total playing and recording time of one hour, consisting of 30 minutes per side. When a half hour of recording time has been used the cassette must be flipped over so the other half hour of recording/playback time is available. The C-60 is the most often used tape, but tapes such as a C-90 and C-120 are also available. A C-90 has a total of 90 minutes of operating time, 45 minutes per side, while a C-120 supplies two hours, one hour per side.

Since a C-60 has the capability of recording 30 minutes on each side, the number of messages that can be recorded depends on their length. A 35-second message is typical and if the messages average this length, then approximately 50 messages could be stored. If side 1 of the tape was used for recording you have the option of rewinding the tape and in that case all previously recorded messages will be erased as the new messages are received. Alternatively, if you want to store the messages you need simply flip the tape to its side 2 (sometimes the letters A and B are used instead of numbers) and then use that side for recording messages. This side will also record about 50 messages of 35 seconds average.

THE ERASE FUNCTION

When the TAD is put in its record mode the tape will be automatically erased to permit new messages to be recorded. If a message tape is full you need not go the trouble of erasing it. However, once a tape is erased it is not possible to recall any messages that have been recorded. You can keep recorded messages indefinitely by removing the cassette and storing it. To hear a message again, put the cassette into the TAD and turn the machine on with its pushbutton control set to playback.

Since the purpose of telephone communications is voice intelligibility only and not entertainment, it is not necessary to use the more expensive high/fidelity tapes. Voice grade tapes are suitable.

It is important to select a quality, brand-name tape for a TAD.

Cheap tapes can cause problems. These include flaking, a process in which magnetic particles fall away from the tape. Flaked areas cannot record or play back. Flaking can also clog the recording/playback head of the TAD, requiring extensive cleaning or an expensive repair.

Another type of tape that is used is the microcassette, advantageous since it is smaller than the standard cassette, permitting the construction of smaller TADs. However, both types of cassettes, standard and microcassette, work in the same way.

A microcassette can provide up to 30 minutes of incoming message time and 15 minutes of outgoing message time. The length of outgoing messages can be variable up to 15 minutes in length. The tape can have two separate outgoing messages with automatic switching when the incoming message tape is full.

COMPATIBILITY

All TADs are compatible with respect to micro or standard cassettes. This means you can use a standard cassette in any answering machine designed to work with such cassettes. Thus, if you have an answering machine in your home and another in your office, and both use standard cassettes, then standard cassettes will fit into both machines. Further, the cassettes can be those made by competing manufacturers.

Microcassettes are also compatible and will fit into any answering machine designed to accept such cassettes. However, microcassettes and standard cassettes are not compatible since they are physically different.

There is no need to remove cassettes from answering machines when the machines are not in use. However, if the cassettes are removed for any reason, keep them away from strong magnetic fields. Thus, for example, do not store them on top of speaker enclosures or near large motors. Both types of cassettes, standard and micro, are susceptible to heat.

SINGLE VS DOUBLE CASSETTE TADS

Some telephone answering devices are single cassette types; others use two cassettes. With a single cassette machine you will be forced to listen to your own recorded message for every telephone call the machine records. It also means you must prerecord your announcement some 20 to 30 times. Since, with a single cassette, you must repeatedly record your announcement you use tape space; the consequence is you have less space left for messages. It is further wasteful of tape space since you have no way of knowing in advance just how much space to leave for messages. For these reasons the dual cassette TAD is superior to the single cassette type.

MESSAGE DELIVERY

You can have the TAD deliver your phone messages to you simply by putting the machine into rewind and then into playback. You can do this whenever you are close enough to the machine to operate its controls.

However, many TADs are now equipped with a Remote Command. Using such a command and a telephone at any distance from the telephone answering machine, the unit can be made to call out each message it has received.

There are two ways in which this can be done: by using a beeper (Fig. 7-4) or by voice (vox) command. The beeper is a compact pocket coder. Its output, a beeping signal, is held near the telephone

Fig. 7-4. Telephone answering machine with beeper-type coder.
(Courtesy Code-A-Phone)

mouthpiece. Any telephone can be used with the beeper, whether rotary or pushbutton type.

When the TAD hears the signal supplied by the beeper it will then begin playing back messages which can be repeated as many times as desired. Since the recorded messages will be of different lengths, the tape will automatically fast-forward to the next channel if the message was short.

The advantage of remote playback is that you can control your telephone answering machine so you can call at any time from the outside and listen to the messages you received while you were away. These outside calls can be made both nationally and internationally. Another advantage of calling in to your telephone answering machine is you can always use direct dialing, the most economical calling service.

The difficulty with telephone answering machines that will respond to a beeper signal is that the user must carry the beeper. This can be a nuisance since, however small, the beeper does require space. It is also possible to forget the unit, lose it, or to find suddenly (and at the wrong time) the batteries need replacement.

Vox operated TADs are "beeperless." The user simply commands the unit by voice, using any type of telephone, to play back messages, rewind, or even change announcements. The answering machine shown in Fig. 7-5 is microprocessor controlled, uses dual minicassettes, has message scan, repeat, two-way recording. It also has a jack to accommodate a cordless telephone for calling screening.

Basically, then, remote control units for telephone answering machines can be either VOX or beeper types. But within these two categories remote control units can have a variety of features. Thus, the remote unit could be a three-button type with controls to direct playback, reset, skip, and repeat messages.

GETTING THE MESSAGE

Every device has limitations and the telephone answering machine is no exception. While it is eminently suitable for supplying a prerecorded announcement to your callers and for taping their replies, those replies will remain on the tape until you direct the TAD to give you your messages. You can do this on premises by depressing the playback button on the machine or remotely by using a beeper, or if a VOX type, by using your voice.

This puts the burden on you of calling your TAD and you may well be put to the trouble of making a number of calls at a time when you have not received any messages. Thus, how rapidly you get your messages depends on how often you telephone in and make contact with your answering machine. In short, once your answering machine has taken your messages, the responsibility for following up on those messages is yours.

Fig. 7-5. Beeperless, microprocessor-controlled TAD *(Courtesy Pathcom, Inc.)*

There are several ways of overcoming this problem and one is by using a machine that will follow through and will try to reach you by phone to alert you about your messages. The component can be pre-programmed to call as many as 12 different telephone numbers. If it gets no response from any of these, it will repeat until contact is made. When you do call in, using a remote coder will give you a replay of the message, an action that causes the component to discontinue trying to follow you any more with further phone calls. The calling device in this answering machine will then automatically reset its preprogrammed list of telephone numbers, ready to use them again upon completion of the next message it receives, and in the same sequence. At any time you want this telephoning action to stop you can turn this function off remotely via telephone.

AUTOMATIC SIGNALING AND PAGING

There is another telephone answering machine that monitors all incoming messages recorded on your answering system and then automatically alerts you about these messages, making it possible for you to respond immediately to your caller's messages.

The component is attached to the same telephone line as your answering machine. When the caller begins speaking, the unit activates and immediately after the message is recorded begins dialing, either rotary or tone, the preprogrammed numbers you have stored in its memory. It only alerts you when a message is left on the recording tape and you will be notified of this message just seconds after it is recorded.

In addition to trying to reach you at the telephone numbers you have designated, the unit can reach you almost anywhere if you are a subscriber to a paging service. It cannot only phone any telephone number up to 20 digits, but will call any type of direct dialing paging service as well.

Since to the unit the paging service is just another telephone number it has been programmed to call upon receipt of a telephone message, you can give the paging service phone number any order of priority you wish. Thus, it can be the first number, somewhere near the middle, or the last. If a voice pager is used the component will transmit a series of identifying tones over the voice pager channel. If it calls a telephone number, it gives out the same identifying tones, letting you know a message is waiting.

Every message recorded activates the unit and it will automatically continue to remind you every four minutes that a message is waiting. However, the component is also equipped with a *Single Call* position which permits it to call you just one time per recorded message.

TAD CONTROLS

The number of controls on a telephone answering device is directly related to the number of features it has, whether the unit is a single or dual cassette type, and whether the unit can be remotely controlled or not.

Fig. 7-6 is a drawing of a TAD that is not remotely controlled. Its OGM (outgoing message tape) is a 20-second endless loop cassette while its ICM (incoming message tape) is a C-60 standard cassette.

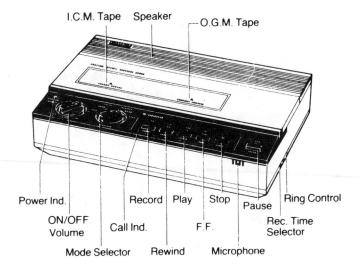

I.C.M. Tape   Speaker   O.G.M. Tape

Power Ind.   Record   Play   Stop   Pause   Ring Control
ON/OFF   Call Ind.   F.F.   Rec. Time
Volume   Selector
Mode Selector   Rewind   Microphone

Fig. 7-6. Controls of a TAD not equipped with a remote device.
(Courtesy Panasonic Co.)

The unit is equipped with cue, review, fast forward (FF), pause, and quick erase. Its ring control is adjustable for rings of 1 to 5 times. It is equipped with a 2.5-inch permanent magnet dynamic speaker and an electret condenser microphone. It also has a modular plug cord that is 7 feet long.

Fig. 7-7 is a drawing of a TAD that can be remotely controlled. The vox remote controller uses three "AAA" size batteries. The OGM is a 20-second endless loop cassette with two channels — one for message request, and one for the anouncement. The ICM is a C-60 regular cassette.

ANSWERING MACHINE FEATURES

Not all TADs are alike and not all have the same number of features. Thus, some do not have a remote call in beeper or VOX. Some are single-cassette types; most use dual cassettes. Some use standard cassettes; others have minicassettes.

The fact that one answering machine has more features than another does not mean it is automatically a better machine, but it usually means a higher price. To determine which answering machine to

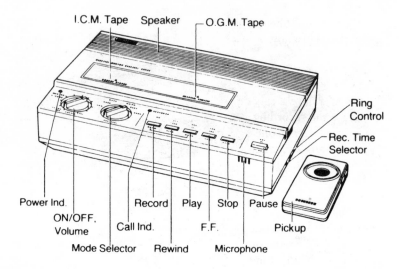

Fig. 7-7. Controls of a TAD using vox remote control. *(Courtesy Panasonic Co.)*

buy, make a list of the features you would like to have and use this list as your purchasing guide.

Limited Voice Actuation

With this feature the voice of the caller signals the answering machine to start taking a message. When the caller stops talking, the machine automatically stops recording.

Automatic Stop

Some answering machines will recognize the fact that a caller has hung up his phone and will stop recording within seconds. This does not happen with all voice-actuated answering machines, and so what you will hear will be a recording of the caller's voice, followed by a recording of the dial tone. The sound of the dial tone may last for just a few or for a number of seconds.

Incoming Call Monitor

You can use your answering machine as your telephone operator when you are at home. The machine will "pick up" the phone for you, not physically, but electronically. You will hear the caller's voice without lifting the receiver. You can then decide whether you will answer the telephone call yourself, overriding the machine, or whether you will allow the machine to record the message for you.

Ring Control

Some TADs have a ring control that lets you adjust the number of rings before the recording machine goes into action. The number of rings can be as few as one or two or as many as seven or more. Many people will hang up, when making a phone call, after the fifth ring. Thus, you can set your machine to answer the telephone, possibly after the fourth ring.

It might seem logical to have the TAD answer the first ring. However, letting the machine wait until a later ring gives you time to decide whether to override the machine and handle the call yourself or to let the machine record it. If you have no intention of answering the telephone at any time, preferring a recording, you can set the device for minimum ringing, possibly one or two.

You can also have variable ring control. Thus, if the TAD can be activated remotely, you might want it to record after the first or second ring when you are away, but to record after the fifth ring when you are at home.

Single- or Two-Way Recording

Many telephone answering machines will simply record incoming messages, but some are available that can record both conversations, incoming and outgoing.

Operating Power

Telephone answering machines are ac operated and so must be connected to an ac power outlet. Most have an on/off power indicator that glows when the power is turned on. The power switch could be a piano key control immediately adjacent to all the other controls. TADs must also be connected to the telephone line and for this purpose they use a modular plug.

UL Listing

When buying an answering machine make sure it is UL (Underwriters' Laboratories) listed. This means it has met standards for fire prevention.

Fast Forward

With this feature the tape can be made to move forward extremely rapidly. This saves time when playing back recorded messages and with its help you can skip over messages that have no interest for you. Quite often just the first few words of a message are all you need to hear to decide if you want to listen to the rest of the message. With fast forward you can move rapidly from one message to the next.

Automatic Shutoff

This function turns the machine off at the completion of message time and also at the end of the tape.

Digital Call Counter

The digital call counter displays, numerically, the total number of calls received by the machine. It is usually a blinking LED (light-emitting diode).

Selective Erase

For most TADs recorded messages are erased sequentially. However, a TAD equipped with Selective Erase can let you remove any recorded message no matter where it is located on the tape. The other messages on the tape are not affected and remain ready for recall.

Message Received Alert

The TAD may be equipped with a flashing red light to indicate when a message has been received.

Sound

TADs are generally equipped with a builtin speaker. This means you need not use a handset to listen to recorded messages.

Disconnect

As indicated earlier, some persons resent being asked to talk to a machine and may hang up without leaving a message. The telephone answering machine, upon hearing a dial tone instead of a voice, acts to disconnect the phone line.

Dictating Machine

Some TADs can perform a double function, working not only as a telephone answering device, but also as a dictating machine. Its usefulness in this respect, though, is quite limited and is governed by available tape recording time. It is suitable for relatively short periods of dictation or for leaving messages for various members of a household. It can serve the same purpose in an office.

Memo Record

You can use the telephone answering machine, not only for receiving and recording incoming telephone calls, but as an electronic memo reminder for members of your family or business associates. Just press the record control and talk up to a maximum of 30 minutes. A light indicator on the unit will glow as an alert that a message has been recorded.

Automatic On

This feature puts the machine in an automatic answering mode should you forget to do so after completing another function with the unit.

Personalized Announcement Recording

You can record your personalized announcement from 2 to 30 seconds by using the builtin condenser microphone. You can always erase the announcement or modify it at any time. While you play back your announcement, you can always learn just what you sound like by calling in from the outside and listening to that announcement. The announcement can be your own voice or the voice of someone you think has a better telephone style.

Glow Light

The machine may be equipped with a Glow Light to alert you that the machine has been turned on, is functioning, and is ready to receive messages. Some telephone answering machines are equipped with a "forget to set" function. This lets the user activate the answering machine simply by calling his own number and letting the phone ring 10 times.

Private Line Use

Some telephone answering devices have a private message channel, enabling one to record an outgoing private message which can be accessed and heard only via the remote control. This is particularly useful for the husband and wife who can leave messages for each other via the answering machine.

Incoming Message Length Control

Some units have a Voice Control Switch giving the user the option of limiting callers to a 45-second message. This compels callers to keep their messages brief. The same switch can be set so that callers can leave long messages, up to 30 minutes maximum.

Variable vs Fixed Recording

Two forms of message recording, variable and fixed, are available. In its variable record mode the unit can record messages so that the machine will stop recording when the caller finishes a message or when the phone is hung up. With fixed recording the telephone answering machine will record calls via fixed time intervals, generally 30 seconds.

Remote Control Coding

With individual coding the user can change the call-in code at will to keep messages secure should the remote control become lost or fall into the wrong hands.

Single Cassette TADs

If the phone answering machine uses just a single cassette the incoming and outgoing messages are contained on different tracks of the single prerecorded cassette. The prerecorded tape could have musical jingles or jingles plus standard voice messages.

Full Message Alert

A tone will sound to warn the user that the incoming message tape is full.

Extension Telephone

Some telephone answering machines are equipped to let you attach an extension telephone.

Announce Only

As indicated earlier, not all answering machine users want to record incoming messages, but simply to alert the caller to try making a call at some other time, possibly specified in the OGM tape. The machine will play your outgoing announcement to the caller and will then hang up automatically.

Time Control vs Voice Control

The TAD can be either voice (vox) or time controlled. In time control the tape receiving the message operates for a predetermined time following the announcement. This represents an economical manufacturing technique, but it has a number of disadvantages. It is wasteful of tape in the sense that it reduces the amount of tape available for useful recording. In some instances, when the message is fairly long, it will cut off the message. In time control the assumption is made that all messages will require the same amount of time.

The other method, voice control, is superior, but the result is a higher initial cost for the machine. With this technique recording stops when the caller stops talking.

Second Announcement

It is possible to lose an incoming message if the tape being used to record messages is full. In some telephone answering machines, when this happens, the devices switches automatically to a second announcement which tells the callers their message is not being recorded, to call back later, or to telephone you at some other number.

The Microphone

Some of the older answering machines made use of the separate microphone. However, nearly all of the more modern units have a built-in electret condenser microphone for announcements or memory recording.

Repeat That Message

Some telephone answering machines will allow you to demand, and get, a message repeat. Thus, you may want to be sure you heard the message correctly, or you may decide you want to write the information the telephone answering machine is supplying. If you are using the machine on premises, getting a repeat is simply a matter of reversing the tape, easily done by depressing the reverse button. But this method is a matter of guessing. With a repeat-message function, you can get tape reversal fairly precisely. Further, you can do this either on premises or via a remote telephone call.

Remote Save or Erase

When calling your answering machine from an outside line, you may want the machine to save all your messages for you. Alternatively, you may want the machine to erase all old messages. Thus, when you call in again, you will not be forced to listen to messages you have heard previously. The machine may have a feature which will let you use your beeper or VOX to save messages or to rewind the tape to its beginning or to erase old messages as new ones are received. If you instruct the machine to rewind the tape to its beginning it will automatically erase as new messages are received.

Announcement Changes

With some machines equipped with remote command, you can change your announcements as often as you wish, for you can just phone them in. You can also review, repeat or skip messages or leave a marker so you won't hear old messages every time you call.

Battery Operation

The remote unit is battery operated and most often uses three AAA size cells.

Outgoing Message

This is generally a 20-second endless loop cassette with two channels: one for announcement, one for message request. On the outgoing tape cassette you can record a message such as "at the sound of the tone please leave your name, telephone number and message and I'll call back as soon as possible."

Incoming Messages

The advantage of using standard cassettes is that you can use them to play back messages on any cassette deck, whether that deck is part of your hi/fi system or is part of your auto sound system. Thus, you can take your message tape along with you and listen to your messages while you drive.

These cassettes operate at 1⅞ inches per second. Elapsed time for fast forward or rewind varies with different machines but a typical example would be 120 seconds for either of these modes.

The incoming message tape is often a C-60 leaderless. This means the tape will record for one-half hour each side.

Generally, audio cassettes have a *leader* at the beginning of the tape. This short length of tape is nonrecording and enables home recordists to make prior adjustments in the controls of the tape deck. This is not needed for TADs so the leader is omitted.

When using a C-60 to record messages, a single message could have a length of one-half hour. This means, of course, no further incoming messages can be recorded on that side of the tape. Or, your announcement could include a suggestion that callers limit themselves to some maximum amount of message time, possibly 30 seconds. Manufacturers often recommend a C-60 for TAD use.

THE TAD TELEPHONE

If the answering machine is integrated with its own telephone, then, in addition to studying the features of the TAD, you should be aware of the features of the phone. Thus, it may be equipped with a last number redial, or may have a number memory in addition to this, or any other features you might like to have.

The advantages of having a TAD that is telephone equipped is its convenience and space-saving characteristic. But it is quite unlikely the TAD and the phone will both have all the features you want. An alternative is to buy a TAD and a telephone separately. In that way you can get these components with all the features you want.

AUTOMATIC SIGNALING AND PAGING

An answering machine will record messages for you and will also respond to an outside beeper or voie command if it is designed to do so. There is another answering machine that monitors all incoming messages recorded on your answering system but then automatically alerts you to these messages, making it possible for you to respond to your callers' messages immediately. For most TADs, though, you will not learn about your messages until you question the machine either by beeper or vox.

With an automatic signaling and paging TAD, the unit activates when the caller begins speaking, but immediately after the message is recorded begins dialing, either rotary or tone, the preprogrammed numbers you have stored in its memory. It only alerts you when a message is left on the recording tape and you will be notified of this message just seconds after it is recorded.

Normally, the unit is programmed to call any type of direct-dial paging service as well as any telephone number of up to 20 digits. If a voice pager is used, the component will transmit a series of identifying tones over the voice pager channel. If it calls a telephone number it gives out the same identifying tones, letting you know a message is waiting.

Every message recorded activates the unit and it will automatically continue to remind you every four minutes a message is waiting. However, the component is also equipped with a "Single Call" position which permits it to call you just one time per recorded message.

TAPE ERASURE

As a magnetic tape passes through an answering machine during the recording process, it is first erased automatically by an erase head. This erasure is immediately prior to recording so the received message has a "clean" magnetic surface. The answering machine will erase messages one at a time, sequentially, eliminating old messages so as to make room for new ones.

You can use an accessory known as a *bulk eraser* for the same purpose. There are some advantages in doing so. The bulk eraser is faster and will erase the complete tape. In some instances the answer-

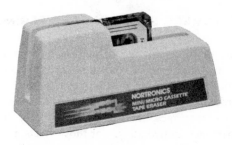

Fig. 7-8. Bulk tape eraser for mini or microcassettes. *(Courtesy Nortronics Company, Inc.)*

ing machine may not do a complete erasing job, and so the remnants of a previously recorded message may interfere with a new one. Also, if you have a tape that seems to have a high level of accompanying noise, using a bulk eraser may help alleviate the difficulty. The unit shown in Fig. 7-8 can be used for either mini or microcassettes. Comparable devices are available for standard cassettes.

WARRANTY

When buying telephone equipment, and that includes telephone answering devices, be sure to keep your sales slip since proof of purchase must be shown for in-warranty service. The warranty generally excludes damage due to abuse, misuse, or using an in-home device for commercial work. Warranties are generally limited to one year.

CHAPTER 8

TELEPHONE SECURITY

Telephone security can mean many things. It could mean you consider the telephone as your first line of defense, a logical and correct approach. Telephone security could also mean privacy in your business conversations via phone. It could mean protection against unauthorized use of your phones, which is equivalent to theft since you will be the one who pays the bills. Telephone security can also mean protection against unwanted calls, whether these are sales solicitations, or obscene. Telephone security means quick help for the bedridden. It means a way of communicating with police or fire departments.

TELEPHONE THEFT

Telephone theft can mean the unauthorized use of your telephone or actual theft of the component itself. It is very easy for a thief to steal any telephone since the telephone plug, standard or modular, connecting the unit to the phone lines is so easy to disconnect, something that can be done, noiselessly, in just a few seconds.

If the telephone is an older model, a rotary type or a pushbutton unit having few features, it is probably safe from theft since the resale value is so low. But some of the more feature-equipped telephones and some of those that are unique can easily have a resale value of several hundred dollars. And since the demand for such units is quite high, the chances of being "ripped off" are good.

The problem of telephone theft is compounded by the fact that some of the newer telephones are very compact and can be easily concealed, are quite light and can be easily carried. Telephone adjuncts, such as answering machines, are now also greatly reduced in size and weight, and so can be carried away quite easily. Cordless telephone systems come into the same category.

This means, then, that your telephone equipment falls into the same category as your television receivers, video cassette recorder, hi/fi equipment, etc. You cannot fasten these in such a way as to prohibit their removal. All you can do is to make your home as burglar proof as possible, making sure to lock doors and windows, use an in-home alarm system, and keep a written record (plus a photo) of all your possessions. It is also advisable to have theft insurance.

UNWANTED TELEPHONE CALLS

The best thing you can do about any unwanted telephone call is to hang up. Just say hello, and if there is no answer, put the telephone back in its cradle. If you answer the telephone and all you hear is silence, then you are the victim of a crackpot call. Don't slam the receiver back into its position. The loud click this produces at the other end will give satisfaction to the caller that he (or she) has managed to get through to you. Do not say, "Who is this?" and certainly do not say it several times. The only answer you should make to a telephone call, any telephone call, is hello. The next move is up to the person who called you. Don't let yourself be tricked into engaging in conversation. A dialog can take place only if you cooperate. If you are the target of repeated obscene or nuisance calls, contact your local telephone utility and inform them of your problem.

WRONG NUMBERS

Never give your telephone number to someone who has obviously called a wrong number. The only way a wrong number can reach you is through someone accidentally dialing your number. But the person who has done that does not know your name, your address or even your correct number. That person is a complete stranger, and while the wrong number call may have been perfectly innocent there is always the possibility it could turn into something unpleasant. The correct answer to such calls is, "You have the wrong number." And then hang up. Anyone making a telephone call and reaching a wrong number can get full credit for the error simply by notifying the operator at the calling end and does not need your telephone number for verification.

KEEPING YOUR TELEPHONE OFF THE HOOK

Some telephone users, in an effort not to be disturbed by the ringing of a telephone, remove the telephone from its hook. When this is done, circuitry at the Central Office of the local phone utility manages to send through a reminder signal that can be as disturbing as a telephone ring. Further, anyone trying to reach a phone that has been disconnected in this way receives a busy signal. They take this as a reminder that you are at home and may keep trying to reach you. And if they have a telephone with an automatic redial feature that telephone will keep trying to get through to you.

If you have a number of telephones in your home they usually operate using a single telephone number. Consequently, if one telephone is kept off the hook this disables all the other telephones in the home, keeping them from being used for making any outside calls.

There are several alternatives. One is to use a light equipped telephone. A flashing light indicates an incoming call, but it is much easier to ignore a light than a ringing phone. As indicated earlier, you should not try to disconnect the phone by removing the modular or standard plug from its jack. Doing so will only result in damage, ultimately, to both the plug and the jack. Instead, you can use an accessory on/off switch for this purpose.

Another alternative is to use a telephone answering machine, described in the preceding chapter. With this device you can receive calls, have the machine answer for you and make a tape recording of the message. You can listen while the message comes in and override the machine if you wish.

TELEPHONE PRIVACY

There are various ways to have telephone privacy. One of the better known methods is to have an unlisted phone. Your local telephone company will supply you with an unlisted number, a number that will not be published in any telephone book.

The number 411 can be dialed by anyone who wants to get number information from the telephone company. But if anyone dials this number and supplies the operator with your name, and possibly your address as well, your telephone number will not be released if it is on the telephone company's NP list.

In the U.S. approximately 82 percent of all telephone numbers are readily available through a listing in a telephone directory and there are about 18 percent that are unlisted. Most of-the unlisted numbers are in the Eastern region of the U.S.

Another method of avoiding unwanted calls is to use a component such as the one illustrated in Fig. 8-1. The telephone shown in the photo is not a part of the unit and sits on a base adjacent to the device.

All telephone calls are first routed through the unit which has an electronic voice. Essentially, it is a special purpose computer which gives the person being called control over his telephone. It provides fully automatic call screening, automatic identification of the person who is calling, and automatic identification of the person being called. The unit will also take messages if it is made to work in conjunction with an answering machine.

Persons who make calls must enter a Personal Access Code when requested to do so by the machine. If the caller replies with the cor-

Fig. 8-1. Telephone access control terminal, shown at the right, gives user control over incoming telephone calls. *(Courtesy International Mobile Machines Corp.)*

rect number that number will be displayed in large numbers on a screen. At the same time the unit will supply one of four different tones. Each of these tones identifies the person who is to receive the call.

The number of the protected telephone is listed in the telephone directory. But anyone who calls the number and does not know the Personal Access Code will not be able to get through and the telephone will not even ring. In this way, nuisance calls, obscene calls, and wrong numbers will not disturb the user of the telephone privacy device.

Further, the user of the device will be able to restrict the distribution of his (or her) Personal Access Code. Authorized callers can be supplied with a card such as the one in Fig. 8-2. The owner can also change the Personal Access Code, make deletions, additions, or changes. Those who receive the Personal Access Card can read the back of the card for instructions. Operating instructions are different for pushbutton and rotary dial telephones.

CHILDREN AND THE TELEPHONE

Some parents are permissive about letting their children answer the telephone and may regard this as cute. However, not everyone thinks this way and some callers can be understandably annoyed when this happens.

But if you do permit your children to answer the telephone, you should train them never to say (in the absence of any or all other members of the family) "nobody's home but me." Instead, caution them not to answer the telephone if no one else is at home. Don't expect your children to match wits with someone who may be a menace to your security.

As soon as a child is old enough encourage him (or her) to remember their telephone number (Fig. 8-3). Keep it simple and omit the area code. If you can do so, make a jingle out of the numbers. Remembering a telephone number is easier for a child than to try to recall a name and address. Don't assume once a child knows a telephone number the knowledge is permanent. Check from time to time and keep reinforcing the child's memory.

As soon as you think your child is old enough, start telephone training. Invite the child to use your extension telephone and encourage the child to join in the conversation. However, make some prior

PRIVEC⊛DE
INSTRUCTION CARD

now has a PriveCode Telephone Access Control Terminal.

CALLING INSTRUCTIONS

1. Dial my phone number. _____
2. Wait for PriveCode to answer.
3. Enter your personal Access Code: _____

INTERNATIONAL MOBILE MACHINES CORPORATION

DETAILED DIALING INSTRUCTIONS

Push-Button Telephones.
1. Dial standard telephone number.
2. Wait for PriveCode to answer and ask for your personal Access Code.
3. Push the correct numbers on your telephone's push buttons.
4. PriveCode will ring and tell me you're on the line.
5. If you make a mistake, push the button marked "#" and start over.
6. If you enter the correct Access Code and PriveCode does not respond, follow rotary dial instructions below.

Rotary-Dial Telephones
1. Dial standard telephone number.
2. Wait for PriveCode to answer and begin its counting sequence: "1, 2, 3, 4," etc.
3. When you hear the first digit you wish to enter, simply repeat that digit into the telephone immediately after it's recited. Speak loudly and crisply.
4. PriveCode will emit a "beep" to let you know the digit has been accepted. PriveCode will then continue reciting digits.
5. To enter the 2nd and 3rd digits, repeat steps 3 and 4.
6. PriveCode will then ring and tell me you're on the line.
7. If you enter an incorrect Access Code, Prive-Code will once again ask you to enter the code and you may start over.

ACCESS CODE: _____
NOTE: If my answering machine is attached, you can leave a message by entering Access Code 1-2-3, or by waiting for my phone to ring 7 times.

Fig. 8-2. Card listing Personal Access Code (left) and detailed dialing instructions. *(Courtesy International Mobile Machines Corp.)*

Fig. 8-3. Teach your children their telephone number.

arrangement with the person you are talking to (a relative or friend) and explain what you are trying to do.

It will help if you buy a telephone toy for your child, one with a rotary dial (if that is the kind you have) or with pushbuttons (if that is what you have). Play a game with your child, and pretend you are answering a call the child has made.

Train the child to identify the 0 on the pad or dial. Put some kind of red marker near the 0 so the child can identify it easily. Be sure to accompany this with an explanation and repeat that explanation often enough until you are sure your child grasps its meaning.

You can get an accessory such as the large number dialing pad shown in Fig. 8-4. This fits right over the telephone pad, but since the numbers are now much larger, enables the child to use it more easily. It is also a good teaching device since it will help the child learn the number system.

How old should a child be to start telephone training? That depends on the child. Some can start at two years, others may need to be a year or two older.

The concept behind all this is that you should look on the telephone not only as a communications method but a device that can supply increased security for all members of your family, even those that are very young.

Fig. 8-4. This accessory slips over pushbutton telephone pad. Large numbers make it easier for children to use.

USING THE TELEPHONE FOR PERSONAL SECURITY

Whether you are an invalid or not, it is always advisable to have a telephone on your night table. True, the ringing of a phone at night can be an alarming sound, and it can be startling since it may well awaken you from your sleep, and it can be frightening even if you have anticipated its ringing.

There are a number of steps you can take to improve this situation. The easiest is to turn the loudness control on the telephone to its minimum setting. If this is still too uncomfortably loud you can muffle the sound with a towel, making certain, though, the telephone is still accessible. You can replace the phone with a model that has a soft chime instead of a bell as the ringer.

To get the maximum security benefits from having a telephone on your night table make sure you have a listing of emergency telephone numbers pasted on the telephone body. These should include emergency numbers such as fire, police, hospital, and possibly the numbers of a friend, a neighbor, or relative. Some areas also have emergency ambulance service and if so that telephone number should be included.

For night table use a pushbutton type of phone is desirable. If you are awakened during the night by what you think may be prowlers, the use of a pushbutton type is not only faster but much quieter than the rotary dial type.

Telephones have an excellent working record. However, you may find it reassuring to lift the receiver before you go to bed at night to hear the reassuring sound of the dial tone. It is your link to the world outside your bedroom.

THE 911 NUMBER

In many areas these three digits (911) represent an emergency calling number. If this number is available make inquiries at your local police station, or your telephone company to determine just what services you can expect by dialing 911. Not all areas use the same techniques in handling emergency calls. In some sections all you will need to do is to dial this number and you will receive assistance automatically. In other places you will need to supply your name, address, and the nature of the emergency.

YOUR CREDIT CARDS

Some security conscious persons keep a record of their credit card numbers as well as the number to call if a credit card (or cards) is lost or stolen. You will find it helpful, under these circumstances, to keep those numbers along with a record of numbers frequently called somewhere near your telephone.

You should also carry these numbers with you, but not in the same wallet in which you keep your credit cards. Thus, you should be able to telephone the number supplied by your credit card company either from your home or from the outside. The faster you make this call the less likely you are to suffer a loss through unauthorized use of your cards.

YOUR DRUGSTORE

It is also helpful to keep the telephone number of your nearest drugstore along with other emergency numbers. In this way you can supply a physician with the number of a pharmacist who can receive telephoned authorization to fill a prescription.

FRIENDS AND RELATIVES

Another phone number you should keep in with all your other emergency numbers is that of a relative or a friend, someone on whom you can depend for help if you should need it. Even though you may have a good memory, don't rely on it in case of an emergency. In a stress situation it is very easy to forget a telephone number, even if it is a number you may use regularly.

IF YOU ARE DISABLED

For persons who are disabled the telephone is more than just a convenience. It can be an absolute necessity. It is often helpful to call or visit the special services department of your local telephone company and to discuss your disability with them. Thus, if you have a hearing problem, a speaking or a walking problem, or any other difficulty, they may be able to suggest telephone accessories which will be of help.

If you have your telephone in a sick room or a bedroom you can get equipment which will convert the ringing of the telephone into a blinking alert light. There are also telephones available, or accessory devices which will let you use single-number dialing so you can reach various emergency numbers quickly. A memory-equipped telephone will do this for you.

You can also get a telephone which does not require your lifting the handset. These are speaker-type phones and, while they can be used by anyone, they are especially useful for persons who are unable to handle a telephone directly.

TELEPHONE CALL THEFT

Although many people don't regard it as such, a telephone is as subject to theft as any other article of value. This does not mean the telephone itself is an object of theft, although some decorator telephones can easily come within this category, but rather the services the telephone offers. There is such a thing as telephone larceny and it comes from the unwarranted use of a telephone, generally in a business, and particularly for long distance calls. Such theft may come from within the organization where use of the telephone for either local or long distance calling is regarded as a right or privilege, often an unwarranted one-way assumption.

In some businesses long distance calling is somewhat controlled by having telephones capable of making such calls restricted to certain executive offices. However, unless such offices are kept locked at night it is possible for anyone on the premises to make long distance calls. Since telephone bills may not be checked carefully, if at all, such phone theft can easily go undetected. A simple method of preventing such theft is to put a lock on the phone, as shown in Fig. 8-5. The lock prevents outgoing calls but does not interfere with those coming in. Unfortunately, this locking method is suitable only for rotary dial telephones and these are gradually being phased out of business.

Fig. 8-5. Keyed lock prevents unauthorized calls on rotary dial telephone.
(Courtesy Taylor Lock Co.)

Another method that is surprisingly simple and effective is to equip the telephone line with an accessory on/off switch (Fig. 8-6) connected between the telephone and the phone plug. Most persons, upon lifting a phone out of its cradle and not hearing a dial tone will immediately assume the phone has been disconnected at a central switchboard and will make no further effort toward making a call.

If the various telephones are controlled by a central switchboard they can be disconnected by the operator at the end of a working day, or the board can be set so no telephone is connected to an outside line. In some businesses just a single phone or two is permitted to remain connected and these are in offices that may be locked and to which only authorized personnel have a key.

TELEPHONE TAPS AND BUGS

One of the difficulties of using a telephone is that it may give you a false sense of security. The fact that you are talking into a telephone with no one else in the room, whether at home or in an office, does supply an impression that your conversation is private. It may be, or it may not be. Tapping a telephone is extremely easy and can be done with devices that are hard to see without some help.

There are two basic ways in which your telephone privacy can be invaded. One of these is through a tap; the other is a bug. A tap is exactly what its name implies — a physical connection to your telephone lines. The tap can be on-premises or off, and can be anywhere

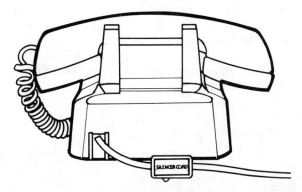

Fig. 8-6. This plug-in phone cord has a handy switch that cuts off the phone. When the switch is turned off there is no dial tone. *(Courtesy Zoom Telephonics, Inc.)*

along the telephone lines. The easiest way to find your lines is before they leave your building.

A bug is any device that can pick up your telephone conversations but without any physical connection to the phone lines. It may consist of a microphone whose output is delivered to a very tiny transmitter. The signal is then broadcast, generally via fm, to a sensitive receiver.

While taps and bugs are often used for telephone conversation theft, this is not always the case. You might, for example, want to tape both sides of a conversation you are having with someone, a perfectly legal process provided the person with whom you are talking has been advised the conversation is being recorded.

Telephone theft involves more than the use of a business telephone for personal reasons. Privileged information is very often communicated by telephone and unauthorized listening to such information puts it right in the class of theft. The word "bug" is used to describe any device which permits listening to one or both sides of a telephone conversation without the knowledge of the parties involved and also without the knowledge of the telephone utility.

There are various telephone devices that will alert you to line-activated devices, taps, wireless transmitters, extension pickups, and line voltage changes.

Using a telephone security alert is easy for all you need do is to plug it into your existing wall jack. It will then warn you, silently, of the presence of a tap on your phone line, of a tape recorder connected across the line, or of a parasitic or self-powered transmitter, and will also alert you to any transmission device that has been positioned somewhere in the area of your telephone. At the same time the unit may be able to deactivate any device on the line, permitting you to continue to use your phone. The component is equipped with an LCD (liquid crystal diode) line voltage readout which permits constant visual monitoring.

If it seems strange that phone calls can be tapped so easily, consider that what you supply to the phone lines when you use the phone is an electrical voltage corresponding to your voice. This voltage can be used to drive a tape recorder. With the help of this recorder the electrical replica of your voice (or that of your caller) is put onto magnetic tape. The unit can then convert these magnetic impressions back into sound.

The voltages corresponding to your voice can also be used to modulate a small transmitter. With this method your voice, in the form of a varying voltage, is carried for some distance where it is picked up by a radio receiver. Your voice, and that of your caller, can then be heard and at the same time put on magnetic tape with the help of a tape recorder.

The simplest way of tapping your telephone line, however, is for someone to pick up an extension telephone at the same time you answer your call. Your voice, and that of your caller, can be heard on any telephone having the same telephone number as the one you are using. The use of an extension telephone can produce a click or some other sound, but since such noises are common, can easily be overlooked.

One way of making certain your telephone conversations are not being monitored is to use a telephone sentry device. The unit detects the presence of one or more telephones being used when you answer your phone.

As mentioned in an earlier chapter, all telephones represent a load across the telephone line, meaning that your telephone requires an operating current, a current that is supplied by your telephone company. The greater the number of telephones being used at any one moment, the larger is the demand for current. It is this increased demand that operates the detecting device. There are a number of different devices available under various trade names.

The component may use one or more light-emitting diodes (LEDs). If one LED begins to glow you have an indication someone is using an extension phone to listen in on your conversation. If two LEDs turn on, then you know two extension phones are being used. The component is battery operated and is equipped with a third LED to supply an indication of battery condition. As long as this third LED glows you know the battery is working. LEDs draw very little current from the battery and, since they are such a light load, can be left on at all times. The unit uses a 9-volt transistor battery and it should have a long operating life.

Connecting the component is easy. It is simply jumpered across the telephone line using a pair of wires. The telephone line wires are color coded green and red. If the unit does not work, transpose the connections to the phone line.

BUGS AND BUGGING

A bug is any device used for surreptitious listening to a telephone conversation. Bugs can be very simple components or highly sophisticated, and some are marvels of miniaturization and electronics. The act of using a bug to listen in on a telephone conversation is called bugging.

At one time radio transmitters used vacuum tubes only and so the size of the transmitter was fairly large. But with the introduction of solid-state devices, such as transistors, integrated circuits, and chips, transmitters were developed that were extremely tiny. Along with these minuscule transmitters were developments in batteries, and these too were miniaturized.

Fig. 8-7 shows a transmitter that is smaller then a quarter. It is capable of transmitting every sound in a room, including a telephone conversation, to an fm radio receiver that may be as much as two miles distant.

This is just one of a family of microtransmitters, small enough to fit into a pen, or into a hollowed out coat button, or a wallet. The sound, picked up by a distant highly sensitive receiver, can be amplified, and retransmitted to any part of the world.

PORTABLE SECURITY

Obtaining telephone security in your home or office is fairly easy. How effective that security is can be another matter since designers of bugs and "bugging technicians" can be highly adroit in both installation and operation. What can you do, then, to assure telephone

Fig. 8-7. Tiny fm transmitter, complete with microphone and battery, is smaller than a quarter. *(Courtesy New Horizons)*

privacy when you are traveling and must use a hotel or motel phone, or a phone in a public place, or a street or building lobby telephone?

One technique is to use the unit shown in Fig. 8-8. It is easy to install, and is a portable device that protects you from unwanted listeners. It is equipped with an alert light to warn you of eavesdropping hotel switchboard or company operators, answering service personnel, outside taps or extension phone listeners. As soon as your call is bugged a red alert light turns on and stays on.

To install the component, remove the telephone mouthpiece and replace it with this telephone guard. There is no interruption of normal phone operation and no wires to connect. The component measures only 1 inch by 2 inches and weighs less than three ounces. It can

Fig. 8-8. Phone-Guard replaces mouthpiece, is easy to install, and assures complete telephone privacy. *(Courtesy Cose Technology Corp.)*

fit in your pocket, luggage or briefcase for use at home, in your office or when you are traveling. The component tells you if the clicking you hear on your line is normal interference of it an operator is verifying or monitoring your conversation. It also alerts you if your telephone is already in use.

TRANSISTOR-OSCILLATOR BUG

There are many types of bugs, but a simple one could consist of a single-transistor operated oscillator. An oscillator is an electronic circuit that generates an ac wave having a fairly high frequency. Known as a carrier wave, its only function is to carry an audio signal, such as that produced by a telephone conversation, for some distance. The operating frequency of the oscillator can be quite high and is generally in the MHz range. If the frequency is between 88 MHz and 107.9 MHz, and if the bug is an fm type, the signals produced by the bug can be picked up and heard by an fm receiver.

Since the function of a bug is illegal to begin with you cannot expect such devices to have FCC approval, nor is there approval for the use of the bug's operating frequencies. The transmission range of a bug is usually quite limited since the purpose of the bug is to pick up a phone conversation, but not to broadcast it over any greater distance than is necessary. Bugging or tapping a phone line is illegal.

Power for the bug can be obtained from a battery and some of these can be incredibly small. Ultraminiature bugs, however, can make use of electrical power obtained from the telephone line. The amount of power taken by the bug is generally so small, since the transistor oscillator requires very little current, that its presence cannot be detected by the telephone company. The voice currents produced by the telephone conversations at both ends modulate the high-frequency carrier wave produced by the transistor oscillator.

What we have, how, are two waves or signals. One of these is that corresponding to the telephone voices; the other is the radio wave produced by the transistor oscillator. The lower-frequency sound waves are then "loaded onto" the higher-frequency carrier or radio wave, a process called modulation. The reason for doing so is that an audio wave can travel only a short distance, something that is not true of the much higher-frequency carrier wave. The concept is comparable to loading cargo onto a truck or freight train. The vehicles are the carriers; the cargo corresponds to the telephone conversation.

The bug, then, is a miniature fm broadcasting station. Its transmitting power is extremely low and since it does not have an antenna, or one that is highly restricted, the distance it can radiate a signal is very small, and is generally within the confines of a house or business.

However, an fm receiver can be a very sensitive device, quite capable of picking up the signals of the bug. When the fm receiver is used it is tuned to the operating frequency of the bug. Both sides of the conversation can be heard, often with the help of headphones. Headphones are desirable in connection with a bug since with their use it becomes more possible to hear everything being said above any electrical noise that may accompany the telephone conversations.

BUILT-IN TELEPHONE SECURITY

It is possible to have a telephone with a built-in security device as shown in Fig. 8-9. You can use the telephone just as you would any other type, but at the same time you are immediately alerted to any line activated device, that is, a bug that turns on automatically when you pick up the phone.

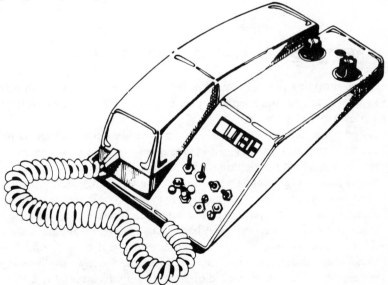

Fig. 8-9. This telephone security device will alert you to line activated devices, extension telephones, and wireless transmitters. *(Courtesy Security Research International)*

The unit will also call your attention to the presence of a wireless transmitter or any changes in telephone line voltage. It is true the telephone line voltage does change when you first start using your telephone, but all during the time you use the phone that voltage remains at a constant level. However, if someone else ties into your line, possibly by using a bug or an extension telephone, then your telephone line voltage changes and becomes lower. There could be an illuminated readout on your telephone that indicates this voltage change.

The component in Fig. 8-9 will alert you, silently, of the presence of a direct wire tap, a tape recorder activator and/or switching circuit, telephone tap transmitters (whether parasitic or self-powered), infinity transmitters, extension off-hook alert, and radio-frequency detection of any transmission device that may have been carried or secreted in your area.

This unit will also instantly deactivate any device on the line thus letting you continue your conversation in privacy. A continuous LCD (liquid crystal diode) telephone line voltage readout allows continual visual monitoring thereby alerting you if significant increases or decreases occur.

PARASITIC VS SELF-POWERED EAVESDROPPING DEVICES

Eavesdropping devices require operating power and can obtain this in two ways: from the telephone line, or from a built-in power source such as a battery.

A parasitic tap is one that uses your telephone line voltage as its source of operating power. These are relatively easy to detect since any change in your telephone line voltage can be indicated on a readout, such as a liquid crystal diode. The advantage of a parasitic tap is that it eliminates the need for a battery. However small, a battery does require space and in a bug the objective is to make the device tiny and inconspicuous. A battery also has a certain amount of shelf-life time, consuming itself whether you use it or not. And, although bugs draw very little current from the battery, the battery, because of its size, is incapable of supplying that current for a long time. Once the battery is weak or dead the tap becomes ineffective, and so there is always the risk of detection when replacing the battery.

ON-LOCATION VS REMOTE BUGS

An on-location bug is any unit housed in the same or a nearby room in which you have your telephone. In a sense your extension telephone is an on-location bug since it can be tapped anywhere that the lines can be reached, whether in or out of your building. The bug may be nothing more than a telephone with leads ending in clips for placement across the phone line. It may or may not be accompanied by a tape recorder so a record can be made of the conversation at both ends.

A remote bug is one that can be operated at some distance from your home or office. It can be attached to your telephone line or it can consist of a microphone secreted in or near your telephone.

The advantage of the microphone is that it is completely independent of your telephone line. There is no way in which the microphone can affect the amount of line current. Accompanying the microphone is an extremely tiny wireless transmitter (Fig. 8-10). Working like a miniature broadcasting station, the microphone picks up your voice and modulates a radio-frequency carrier. This carrier, operating on a single fixed high frequency, is transmitted and is picked up by a remote radio receiver. At the radio receiver your conversation can be heard via earphones or a speaker and your voice can be recorded at that time.

Its independence of the telephone is the main advantage of this device. But it can pick up only one side of the conversation. Further, the battery needs replacement from time to time. The bug, extremely small to escape detection, is fitted into the mouthpiece of the telephone.

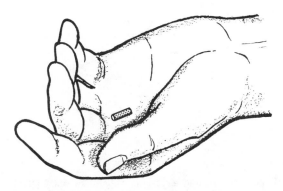

Fig. 8-10. Approximate size of a bug. Some are even smaller.

TELEPHONE SECURITY DEVICES

One of the advantages of a telephone security device as compared to a bug is that the former can be independent of battery size, can be portable or a fixed position component, and may be out in the open or invisible, just as the user chooses.

There are a number of telephone security devices. The one shown in Fig. 8-11 is so small it can be carried in the palm of your hand. It detects and alerts you to the existence of a bug concealed on someone in your presence or planted in the room. It has a tiny signal light which is easily concealed from view, can find electronic bugs in just seconds, and verifies the presence of an eavesdropping device. You can use it anytime, anywhere. It lets you do a quick electronic sweep every time you enter a room.

Just as a bug can be planted in your office, so too can a bug alert be concealed in some ordinary object that doesn't attract attention. Thus, you can get a bug detector that is encased in a working desktop pen set. This will alert you to the existence of a concealed bug that is transmitting your conversation to an outside source. Because it is hidden in the desk set it can be openly displayed in your home, office, or conference room for continuous assurance of privacy.

Another innocent looking protection device is a wiretap trap. It looks and works like an ordinary phone, but mounted in its base is an electronic privacy system that protects against most wiretapping devices.

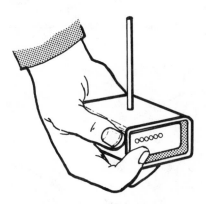

Fig. 8-11. This device can find electronic bugs in seconds. *(Courtesy CCS Communication Control, Inc.)*

There is also available a portable bug alert that is small enough to fit in your pocket. It lets you know you are being bugged by vibrating softly. It does not make any noise and it doesn't use any warning lights, but when you feel the vibrations (which are inaudible) you will know your telephone is being bugged.

Another type of bug detector is concealed in an ordinary pen. You can wear it in your pocket and it will alert you to the presence of a bug by flashing a light.

TAPE RECORDER DETECTOR

Bugs function in a variety of ways. The audio signal produced by the bug can be listened to directly or it can be transmitted so it can be heard off premises. Associated with the bug may be an on-premises tape recorder. To counter this bugging action you can get a tape recorder bug detector. The unit, no larger than a pack of cigarettes, electronically detects any tape recorders in your presence. A series of lights warn you that your conversation is being recorded surreptitiously.

The war between bugs and bug detectors is undeclared, but exists. Many of the bugs being used are highly ingenious, but so are bug detectors. Thus, you can get a bug detector that is mounted inside a wristwatch. This bug detector's output signal can be fed into a pocket type tape recorder detector.

SURVEILLANCE SYSTEM

You can get a surveillance system that will not only find bugs associated with your telephone system but anywhere on your premises. One such unit performs a professional electronic sweep. Intended to be completely inconspicuous, it looks just like a typical radio receiver.

Not all bug detectors are camouflaged. Some are right out in the open, possibly as a warning to visitors that a debugging device is not only present but is active. One of these systems detects, locates and verifies hidden bugs. Furthermore, it lets you tune into the transmitter being used by the bugging device.

ANTITAPPING

If your telephone line has been tapped you can get an antitapping unit that will detect listening devices on your telephone line or equipment. It is equipped with a line analyzer jack to test for both on-hook and off-hook taps. It can also be linked to a tape recorder to give you a permanent record of conversations on your phone that have been tapped.

VOICELESS COMMUNICATIONS

You can communicate without using your voice, so in effect you are "speaking" in total silence. In the unit shown in Fig. 8-12 communication is established by tapping out your message on the equiva-

Fig. 8-12. Voiceless transmission using pressure sensitive typewriter keyboard. Message is displayed on video screen shown at the right. *(Courtesy CCS Communication Control, Inc.)*

lent of a typewriter keyboard. The message appears on the screen shown at the right. The party to whom you are "talking" has a comparable system, also silent, with the message appearing on the receiving screen. The system is duplex. The telephone, positioned as indicated in the photo, receives and transmits messages.

This communications system is contained in a dispatch case, so you can carry it with you wherever you wish. The telephone is easily connected to any modular telephone jack.

The system shown in Fig. 8-12 keeps every telephone conversation free from wiretapping, from tape recording, and free from room bugs that may be hidden on the premises. The keyboard used is a pressure sensitive type so it is completely silent, with no mechanical elements to get out of order. A computer in the unit stores an infinite number of unbeatable scrambler codes for your protection.

Simultaneously, it encodes your message and transmits it over ordinary telephone lines. The intended party (and only the intended party) receives a video display of the conversation on a miniature computer screen. Room transmitters and tape recorders are powerless. Interception of any kind is virtually impossible. At your command, a pushbutton control will convert the visual conversation into audible electronic speech.

The display uses a 5-inch screen capable of showing 24 lines of 40 characters. The power supply can be 120 volts ac or 240 volts, 50/60 Hz. For portability it contains rechargeable Nicad batteries. The unit can be adapted for video color output or printer output.

Another device, along similar lines, uses a pad so you can write your messages instead of typing them. It transmits your messages in total silence using ordinary telephone lines.

SCRAMBLERS

For every device you could use to alert you to the presence of a telephone tap, somewhere, somehow, someone will devise a tap that will escape detection, at least for some period of time. This battle between detection and nondetection is a constant one. For that reason, some companies, to avoid corporate spying and the theft of business ideas, make use of a scrambler. While the telephone is our most convenient and most reliable form of telecommunications, it also represents, possibly, the largest business secret leakage source.

Scramblers can be portable or a fixed-position device, and is a multiple coded, self-contained unit for ensuring privacy between a pair or more of users. A scrambler, more technically known as an encoder, makes normal speech unintelligible.

One way of doing this is by transposing low and high voice frequencies prior to transmission via telephone lines. At the receiving end, an unscrambler, also known as a decoder, inverts the voice frequencies so they are heard in the order they had prior to the encoding process.

Instead of a single scrambling technique, the encoder can supply a number of different scrambling codes. Thus, to ensure maximum privacy, the scrambling codes can be changed each day on a random basis. One portable telephone scrambler can supply 25 different scrambling codes, selectable by a control on the front panel of the instrument. The equipment can also operate, through repeater stations in a base-to-base mode, mobile-to-mobile, or mobile-to-base.

Scramblers have various degrees of electronic sophistication. Given enough time, a tape recording obtained surreptitiously of an encoded voice transmission could possibly be decoded. Whether such effort is economically worthwhile depends on the commercial value of your telephone conversations. The sale of commercial secrets is theft, just as much as breaking and entering. Telephone theft is difficult to prove since there is often no physical evidence, and if evidence is found, difficult to connect to any persons guilty of this crime.

SCRAMBLING METHODS

One method of scrambling, mentioned earlier, is to invert all voice frequencies. Another is to use a tone generator within the scrambler, making possible a number of different scrambling codes by varying the pitch of a translation tone generator. One portable telephone scrambler provides individual speaking and receiving selection of 25 different translation codes on the control panel of the instrument. The letter-coded switch controls the speaking translational tone while a numerically coded switch controls the receiving tone translation.

PERSONAL HEALTH SUPPORT SYSTEM

The problem with the dedicated telephone line described previously is that it assumes the person using the telephone is in its immediate vicinity, but that is not always the case. Thus, the dedicated telephone line limits the individual's mobility.

To overcome this restriction it is possible to have a *Personal Health Support System.* This system consists of a device that resembles a wristwatch, known as a Signal Bracelet, plus a Signal Relay. The telephone, preferably one that is dedicated, is positioned on top of the Signal Relay unit. Fig. 8-13 shows the Signal Bracelet and the Signal Relay.

When the person using the signal bracelet is in distress or recognizes the signs of an oncoming requirement for help, he (or she) simply depresses two small buttons on the Signal Bracelet. This operation sends a radio signal to the Signal Relay unit and the system is then immediately activated.

The Signal Bracelet contains a small transmitter having a range of 150 feet in free air. The bracelet uses a 6-volt battery with an expected life of 6 months. It is recommended that the battery be replaced after five triggerings. The bracelet with its batteries weighs 65 grams. The Signal Relay is ac operated, works from the 120-volt ac, 60-Hz power line and requires only 15 watts of power. It contains an fm receiver, is a single-channel type, and is microprocessor controlled.

When it receives a signal from the Signal Braclet, the Signal Relay can dial up to four preselected telephone numbers sequentially. It will then request help by playing a message you have prerecorded.

The Signal Relay, upon completion of the first call, will disengage and await a confirming call from the first respondent. Should the first number not reply, possibly because the line is busy or there is no

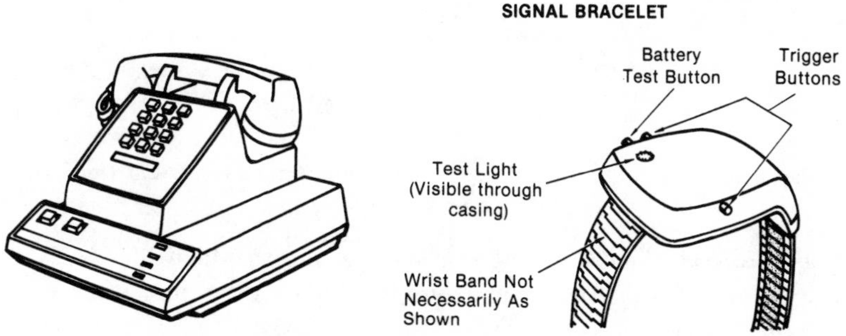

SIGNAL BRACELET

Battery Test Button

Trigger Buttons

Test Light (Visible through casing)

Wrist Band Not Necessarily As Shown

Fig. 8-13. Personal Health Support System. Dedicated telephone sits in space provided in the Signal Relay (left). Signal bracelet can be used at distance up to 150 feet from the Signal Relay. *(Courtesy Almicro Electronics, Inc.)*

answer, the Signal Relay will automatically dial the second number in the sequence and play the preprogrammed message. A tone command on the second number will delete it from the calling sequence and the unit will return to calling the first number.

If both the first and/or second numbers are not reached after repeated calls, the third and fourth numbers will be called. The Signal Relay will then redial the first number until successful. After completing all calls, a confirmation light on the unit will turn on and the calling sequence will be terminated.

TELEPHONE PICKUP

While a telephone answering machine can be used to record both sides of a telephone conversation, it is not necessary to buy the unit for this purpose. If you own a tape recorder you can use the device shown in Fig. 8-14. Known as a telephone pickup, it is very simple and easy to use. It consists of a suction cup, a connecting cord and a jack for plugging into the input of the tape recorder. The component does not require batteries nor is it connected to the ac line. In operation, the rim of the suction cup is moistened and is then securely attached close to the mouthpiece of the telephone. It does not have an on/off switch nor is one required. It can record both sides of a conversation, but your voice will sound much stronger than the reply. The connecting plug is a miniature 3.5mm phone plug and will fit most tape recorders.

Fig. 8-14. Suction cup type telephone pickup. *(Courtesy GC Electronics)*

Before recording someone else's voice, inform them of what you are doing. This device is not equipped with a beep alert often used for this purpose.

Another type of telephone pickup is shown in Fig. 8-15. Instead of using a suction cup it is attached to the telephone earpiece. Like the suction cup unit, it is equipped with a 3.5mm phone plug for connection to a tape recorder.

Fig. 8-15. Telephone pickup slips over earpiece of phone.
(Courtesy GC Electronics)

WALL BUG

The same basic idea of the suction cup pickup can be used as a wall bug. The device (shown in Fig. 8-16) consists of a suction cup, an audio amplifier that is battery operated, and an in-ear phone. The suction cup is applied to the wall of a room in which a telephone is being used. Walls are good conductors of sound and so the wall bug is quite effective. The person using the wall bug can either listen to the conversation, can tape it with the help of a battery-operated tape recorder, or can do both.

The problem with detecting this kind of bug is that it has no connection to telephones lines — that is, it is not a tapping device. Its output is an audio signal, but such signals have a very small traveling range in space. However, if a tape recorder is used in conjunction with the wall bug, the chances of detection are much better. Tape

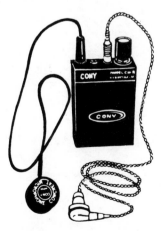

Fig. 8-16. Wall bug is simple and effective device. *(Courtesy New Horizons)*

recorders use a high-frequency bias oscillator, and so the signal produced by this tape recorder circuit can be detected.

CORDLESS TELEPHONE SECURITY

Some cordless telephones are equipped with a security feature, a coding arrangement which prevents the use of your base station telephone by unauthorized persons. Thus, a neighbor with a handset could easily bug your base station if it was within range. Or, a handset could be used to listen to what you had assumed to be a private conversation. This was described in more detail in Chapter 5.

CHAPTER 9

TELEPHONE TECHNOLOGY

Telephone technology, both in telephone components and accessories, and in the transmission and reception of sound, is moving ahead at a high rate. The world of telephones is changing, so much so that even recent telephones and our methods of telephone communications are obsolescent. The use of the telephone has kept pace, for in the most modest home a telephone is regarded as a necessity.

THE ANALOG SIGNAL

A sound, such as that of the human voice, can be converted to a voltage, and since the voice varies in strength, the voltage will change accordingly. A voltage waveform corresponding to single-tone sound is shown in Fig. 9-1. Since the voltage changes in step with the sound producing it we can say it is comparable or analogous to it. Thus, the voltage produced when you speak into a telephone is an analog of the sounds you make. The device for producing this corresponding voltage change is a microphone and is a part of every telephone.

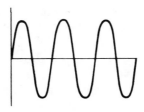

Fig. 9-1. Voltage waveform corresponding to a sound.

263

The microphone is used for changing sound into a voltage since we cannot amplify sound directly. There are some simple devices that concentrate sound, such as a megaphone, but its application is extremely limited. The advantage of converting sound to a voltage is that we can amplify voltages and transmit them through the air (as in broadcasting) or via wires (as in telephony).

MEASURING A VOLTAGE

The voltage in Fig. 9-1 is a varying one with its strength changing from moment to moment. If we used a meter to measure this voltage the pointer of the instrument would swing back and forth violently in an effort to keep up with the changes.

As an alternative we could average the voltage using the approach shown in Fig. 9-2. Here we have a number of vertical lines with the height of each line giving us the amount of voltage at any particular moment. As an example, this series of voltage measurements could result in a group of numbers such as 15, 19, 23, 15, 21, 25, 17, 28, etc. By adding these and then dividing the result by the total number of measurements we could arrive at an average value.

VOLTAGES BY THE NUMBERS

Instead of averaging the numbers representing the instantaneous values of voltage, we could simply list the numbers themselves. We would then have a series of numbers similar to those indicated in the preceding paragraph. The question, then, is what we could do with these numbers and how we could utilize them in a telephone system.

THE DECIMAL NUMBER SYSTEM

These numbers representing the voltage from moment to moment are symbols and are part of our decimal system. This system consists of 10 symbols, ranging from 0 through 9. Any decimal number, no matter how large, is just their combination. A number such as 69,187,205 is a much larger number than 325, but both numbers use the same symbols, although one number, because it is larger, uses more of the symbols. The decimal system is just one of a large group

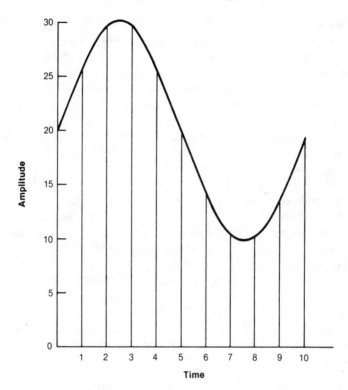

Fig. 9-2. Each vertical line represents the amount of voltage measured at a particular moment.

of numbering systems, but it is the most familiar one since we use it every day.

THE BINARY SYSTEM

A binary system is a system of working with numbers, just as the decimal system is a way of working with numbers. Instead of ten symbols, as in the decimal system, the binary is much simpler and uses just two symbols — 0 and 1. Symbols such as 2, 3, 4, 5 etc. simply do not exist in the binary system.

WHY THE BINARY SYSTEM?

The binary system is extremely important in telephony and, of course, is the system used by computers.

In Fig. 9-3 we have a number of switches identified as SW1, SW2, etc. While there are some switches that are quite complicated, these are simple. They have only two possible working conditions — either they are on or they are off. We can also say they are open to closed.

Now you can see the advantage of the binary system. We can represent the symbols in this system by open and closed switches. We can use an open switch to represent the number 0. A closed switch could then be used to indicate the number 1. Thus, if we have enough switches we can have them represent any binary number.

Although switches are quite ordinary we are not restricted to using them only. A relay is another common device and many of them are used in telephone work. The problem with relays, and it is the same problem we have with switches, is that they are mechanical devices and so are not only relatively slow, but, because they have moving parts, are subject to wear. Fortunately, there are other devices, such as solid-state components, that can also work in an on/off manner. Thus, a semiconductor diode or a transistor, when conducting current, is comparable to a closed switch, and in that case could represent the binary digit 1. By stopping the flow of current through the semiconductor, it could represent digit 0. Semiconductors have the further advantage of being extremely small, lightweight, and without moving parts.

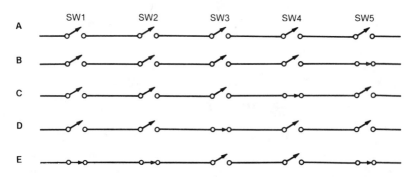

Fig. 9-3. Switches can be used to produce the eqivalent of a binary number. An open switch is 0; a closed switch is 1.

BINARY VS DECIMAL

We can read decimal numbers because we think and work with such numbers. We get paid in decimal numbers, we do our buying using them, and we literally live with them. However, even decimal numbers can sometimes be confusing. Thus, if we have a number such as 937654101 the number is difficult to read and so we generally use commas or spaces to separate the digits. In this example we would have 937,654,101 or we could have 937 654 101.

We have the same problem with binary numbers. A group of binary numbers could be written as 101001011110. For binary numbers we do not use commas but put spaces in their place. And so this number would become 1010 0101 1110. The value of the number hasn't changed, but it has become easier to read.

BINARY TO DECIMAL

Right now these binary numbers are meaningless. For the numbers to have any significance for us we must convert them to decimal form.

Table 9-1 supplies a table of binary numbers and shows their equivalent value in decimal terms. The lowest value binary is the

Table 9-1. Decimal Values and Their Corresponding Binary Numbers

Decimal Value	Binary Value				
	16	8	4	2	1
0	0	0	0	0	0
1	0	0	0	0	1
2	0	0	0	1	0
3	0	0	0	1	7
4	0	0	1	0	0
5	0	0	1	0	1
6	0	0	1	1	0
7	0	0	1	1	1
8	0	1	0	0	0
99	0	1	0	0	1
10	0	1	0	1	0
11	0	1	0	1	1
12	0	1	1	0	0
13	0	1	1	0	1
14	0	1	1	1	0
15	0	1	1	1	1
16	1	0	0	0	0
17	1	0	0	0	1
18	1	0	0	1	0
19	1	0	0	1	1
20	1	0	1	0	0

rightmost digit. This has a decimal equivalent value of 1. The binary to its immediate left has a value of 2, that adjacent to it, still moving to the left, is 4, then 8, then 16. Note that the decimal equivalent value of each binary number doubles as we move from right to left. The decimal equivalents, starting at the right and moving to the left are 1, 2, 4, 8, 16, 32, 64, 128, 256, 510, etc. With this information on hand we can convert any binary number to its decimal equivalent. See Table 9-2.

Table 9-2. The Value of a Binary Number, in Decimal Terms, Depends on Its Horizontal Location. The Lowest Value Binary Is at the Extreme Right. The Value of a Binary Number Doubles for Each Position It Takes, Moving Toward the Left.

1024	512	256	128	64	32	16	8	4	2	1
1	1	1	1	1	1	1	1	1	1	1

Chart 9-1 shows some binary numbers and their decimal equivalents. This means we can express any binary number in decimal form, or, going the other way, show any decimal number in binary.

Chart 9-1. Decimal Numbers and Their Values in Binary Form

DECIMAL NUMBERS	BINARY NUMBERS
19	1 0 0 1 1
41	1 0 1 0 0 1
78	1 0 0 1 1 1 0
20	1 0 1 0 0
15	0 1 1 1 1

EXAMPLE:

         1   0   0   1   1   1   0   binary
64 + 0 + 0 + 8 + 4 + 2 + 0 = 78   decimal
                                          equivalent

PULSE FORMATION

Fig. 9-4 shows a single circuit consisting of a battery, a single-pole, single-throw switch and a lamp. When the switch is closed, current flows practically instantaneously through the lamp, and when the switch is opened, current drops to zero just as quickly.

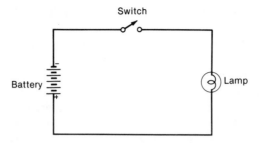

Fig. 9-4. Pulses of current are produced every time the switch is opened and closed.

We can graph the behavior of the current as shown in Fig. 9-5 (also shown earlier in Chapter 1). The bottom horizontal line represents the passage of time. When the switch is closed the current reaches its maximum, as indicated by the vertical line. The current then flows for as long as the switch remains closed, and this is shown by the horizontal line across the top. If we then open the switch, the current drops to zero. This is indicated by the second vertical line.

What we have here is a graphic representation of a single pulse of current, but by operating the switch we can have a number of such pulses. These pulses, though, can represent binary numbers. An on pulse is the same as binary 1; an off pulse is the same as binary 0. Hence, we can use a series of pulses to represent binary numbers.

But these binary numbers are the equivalent of decimal numbers. In turn, these decimal numbers were taken from the analog waveform representing the voice. And so, by this chain of events we

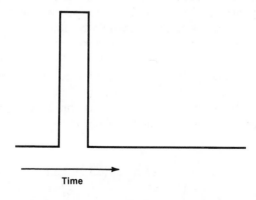

Time

Fig. 9-5. A single pulse of current can represent binary digit 1. The absence of a pulse is equivalent to binary digit 0.

have a series of binary numbers which are the equivalent of voice sounds.

The result of all this is that instead of sending the analog of the voice through telephone lines we transmit pulses instead. At the entrance to the telephone line we need an analog-to-digital converter, a device that will automatically change the analog waveform to pulses. It does this by extremely rapid sampling of the wave. At the other end of the line, before those pulses are sent to the person receiving the phone call, is a digital-to-analog converter. This device changes the pulses of current back to their analog form, and that is the electrical equivalent of the voice.

Consequently, in digital telephony, we analyze the electrical waveform corresponding to the human voice, and from it derive its equivalent in digital form. We then transmit these pulses, and finally change back to the analog wave, our representation of the voice.

Digital transmission has a number of advantages. It supplies telephone communications much less susceptible to noise. Further, digital transmission requires far less room in the connecting cables, and so many more signals can be sent along the same wire. This increases the capacity of the telephone system.

THE CONCEPT OF BANDWIDTH

A complex tone can be represented as a wave, shown in Fig. 9-6. This tone could have a particular pitch or frequency, measured in cycles per second or hertz (Hz). Thus, the wave in the drawing could have a frequency of 50 Hz, or 50 cycles per second (cps). Another wave having twice the frequency would be 100 Hz.

Unlike these single-frequency waves the human voice consists of a number of frequencies, ranging approximately from 200 Hz to 3000 Hz, with the range of frequencies determined by a number of variables such as sex, age, and physical condition.

Consequently, when we speak into a telephone we supply that instrument not with sound of one frequency but many. The range from

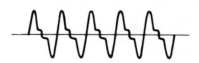

Fig. 9-6. Waveform corresponding to a complex tone.

the lowest to the highest is called the bandwidth. The bandwidth of the human voice is about 3000 Hz.

Of course the telephone lines must be able to accommodate the human voice, somewhat of a problem since the voice consists of not one signal or tone, but thousands. If these tones could be restricted the telephone lines could accommodate many more of them. And that is precisely what digital telephony accomplishes. Instead of transmitting the voice and its fairly wide bandwidth through telephone wires, the voice is changed to its corresponding pulse or digital form, and this is practically equivalent to a single-frequency tone.

PULSE FREQUENCY

The simple circuit shown earlier in Fig. 9-4 produces pulses by means of a switch. While the switch could be operated manually this would be impractical and so such switches are controlled electronically. This means we can produce uniform pulses and do so extremely rapidly. With such a switching arrangement we can also determine the number of pulses we want to produce per unit of tme. Thus, we could have 5 pulses per second (pps), or 10, or 100, or 1000 or more.

BINARY DIGITS

As indicated earlier, the binary numbering system uses two symbols only, 0 and 1. Each of these numbers is a binary digit, more often described by an acronym, *bit*, a condensation of *b*inary dig *it*.

DIGITAL TRANSMISSION

Ordinarily we think of telephones as a device for voice communications but telephone lines and broadcast transmitting systems can be used for data. Data means numbers, but numbers in this case refers to binary. Not only numbers, but letters as well, can be converted into binary form, and so the transmission of data is alphanumeric.

To transmit data, letters and decimal numbers are encoded into binary form. The data is then transmitted and at the receiving end a decoder puts the binary information back into decimal form or letters, or both.

Fig. 9-7 shows one way in which digital data is handled. At one end the information is collected and arranged in the usual language form. It is then converted to bit form (binary digits) by an encoder, a device for changing analog information to binary. The signal can then be transmitted via telephone lines or by microwaves. A digital decoder at the receiving end converts the binary digits into decimal numbers, letters, or both.

This method of transferring information is a one-way process, hence is simplex (SX). By duplicating all the components it could be converted to duplex as shown in Fig. 9-8. One of the units shown in this illustration is a modem, an acronym for *mod*ulator-*dem*odulator. A modulator is a device for loading a signal onto a higher frequency carrier wave. The purpose of a carrier wave is exactly what its name implies — to carry a signal, in this case the digital data. Broadcast stations, am, fm, and tv all use carrier waves to "carry" information to receiving equipment in your home. Once the delivery has been accomplished, your radio receiver or tv set discards the carrier. The sound is sent along to a speaker; the picture information to a picture tube.

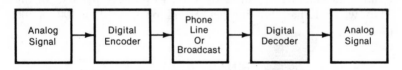

Fig. 9-7. Data transmission method.

MODEM

A modem is a device that can either modulate a signal (load it onto a higher frequency carrier wave) or demodulate it (remove the signal from its carrier wave). See Fig. 9-9. The process of modulation is always associated with the transmission of a signal; demodulation with the reception of a signal.

While the usual telephone (there are some receive-only telephones) is a duplex device, it does not require a modem since the transmission is at audio frequencies over wires.

In Fig. 9-8 the encoded data is supplied to a modem and then is sent along telephone wires or through space and is then picked up by a receiver. Another modem is used, this tme to separate the data from the carrier wave. This process is the opposite of modulation, hence is

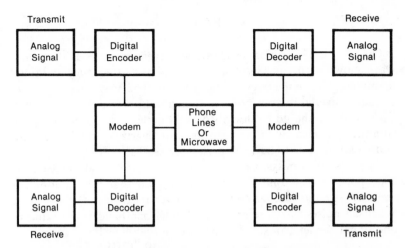

Fig. 9-8. Duplex digital transmission system.

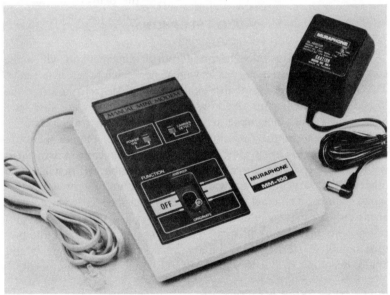

Fig. 9-9. Mini-modem sends and receives information at 0—300 baud.
(Courtesy Mura Corp.)

called demodulation. Following demodulation, the data is fed into a digital-to-analog converter or decoder. The information can then be viewed on a video screen, or can be printed out, or both. The arrangement in Fig. 9-8 is duplex.

BAUD

One way of describing an electrical waveform, such as the sound wave shown earlier in Fig. 9-6, is in terms of its frequency and time, or number of complete cycles per second. The frequency of the ac voltage delivered to your home is sometimes referred to as 60 cycles. Technically, it should be described as 60 cycles per second. To make sure the time element is included the word hertz is now being used, and so 60 hertz (60 Hz) is 60 cycles per second.

For pulse transmission the word *baud* is used. 1000 baud means 1000 pulses per second with the assumption that these pulses are identical. A baud is usually equal to one binary digit (bit) per second. When all pulses have the same amplitude, one baud is the same as one signal element, or one bit, per second.

VIDEO TELEPHONY

The concept of a video telephone, an arrangement in which the communicating parties can see as well as hear each other, is not new. The Bell Telephone Company exhibited a video telephone at the World's Fair in Flushing Meadows, N.Y. in 1939. They are also demonstrating a video telephone at Epcot Center in Central Florida.

While it is technically possible to have video telephones there are several obstacles. One is the cost of having a video camera associated with each telephone, plus a monitor picture tube for viewing. The other is the very high bandwidth required. Our present day video signal requires a channel that is 4,200,000 cycles wide. A voice signal needs only a range from about 200 to 3000 cycles. One possible solution to the problem of adequate channel space is to encode the picture signal, converting it to binary form, then transmitting it and finally decoding it. But this requires additional, costly equipment.

Video Door Telephone System

Ordinarily, a video telephone is regarded as a supplement to voice communications, but the telephone plus a video accessory can also be used for security. Unlike the video telephone the video door telephone is presently available.

Instead of the usual doorbell the door telephone uses a tv monitor (Fig. 9-10). When your visitor presses the call button you will hear a

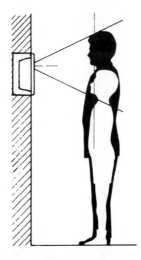

Fig. 9-10. Outdoor video camera system includes lights, a speaker, microphone and a call button. *(Courtesy Unisonic Products Corp.)*

chime. Outdoors, the lights go on and the camera built into the outdoor unit will begin to record. If you do not answer, the lights and the camera will turn off automatically after 30 seconds.

On the inside of your home the picture of your visitor will appear automatically on a tv monitor. To talk to your visitor all you will need to do will be to lift the handset of the accompanying telephone and talk into it. Your visitor will hear your voice in the speaker and will be able to respond by talking into the microphone of the outdoor camera unit (Fig. 9-11).

Fig. 9-11. With the indoor video telephone you are able to see as well as hear your visitor. *(Courtesy Unisonic Products Corp.)*

OPTICAL FIBERS

Telephone lines can be compared to water pipes. In both instances we want the lines or pipes to carry as much as possible. The fewer the number of telephone lines required, the more economical is the transmission of signals.

This problem can be approached in two ways. One is through the use of digital transmission as explained earlier. The other is to use materials other than copper wire to carry the signals. One of these copper-wire substitutes, oddly enough, consists of optical fibers. An optical fiber, as its name suggests, carries light rather than electrical currents.

When optical fibers are used a new method of signalling is required. Thus, a solid-state laser at one end of each fiber generates pulses of light in the long-wavelength portion of the light spectrum. The pulses are detected at the receiving end by a solid-state photodiode designed to respond to long-wavelength light. The fibers carry conversations by working in pairs, one strand for each direction of speech.

If these seem strange, consider that we can use pulses for the transmission of information. It makes no difference whether those pulses are in electrical or light form, since the pulses are a digital representation of an analog signal.

Lasers

The light produced by a laser is used. Ordinary light such as that produced by a flashlight, electric light bulb, or any similar light source, consists of not just one but numerous light frequencies. In that respect it is comparable to the human voice, also with its frequency range. The word *laser* is an acronym, coined from the phrase *l*ight *a*mplification by *s*timulated *e*mission of *r*adiation. A laser is a device for generating radiation in the visible, ultraviolet or infrared regions of the light spectrum. This radiation is called coherent since the light consists of energy of one frequency.

Lasers and Telephony

Laser devices used in telephone work are capable of producing nearly 45,000,000 pulses a second; thus enabling a single pair of glass "wires" to carry 672 phone calls at one time. Two other fibers within the cable are available to handle increased calling volumes in the future. The remaining pair is held in reserve for emergencies.

Use of the second pair of fibers would increase the system's capacity to 1344 simultaneous phone calls. In addition the system can be upgraded to handle 90 million pulses of light a second, thereby doubling its capacity to 2688 calls.

One of the problems with transmitting laser light pulses along a glass fiber is that the light gradually loses energy, becomes weaker. The same is true of electrical pulses along a wire. At intervals along the wire *repeaters* (amplifiers) must be put into the line to reinforce the signals. Similarly, one of the problems with optical cables has been the need for installing regenerators along optical cable routes to rebuild the light pulses before they fade. But this problem has been minimized, although not eliminated, through the development of so-called *superfibers*. These are contained in a flexible pencil-thin cable.

Most optical telephone links presently in service carry calls on short-wavelength pulses which begin to weaken and lose their shape after traveling about 10 kilometers (6.2 miles). This means regenerators must be installed at points along the cable to rebuild the light pulses before they fade on short wavelength routes of more than 10 kilometers.

However, a new system developed by General Telephone Electronics has long wavelength pulses that can travel at least 35 kilometers before they have to be regenerated. This is because light in the longer wavelengths is less susceptible to absorption by the glass fibers than light at the shorter wavelengths.

Long wavelength pulses are in the part of the light spectrum where the light waves are 1300 nanometers long, compared to 850 nanometers for the shorter wavelength system. A nanometer is one billionth of a meter which is 39.4 inches long. Both the long and short wavelength pulses are in the infrared portion of the light spectrum and are therefore invisible to the human eye.

Optical systems can not only carry a large volume of communications signals, but can also provide reliable clear circuits by virtue of the fibers' immunity to electrical interference, lightning, and *crosstalk* from other channels. Crosstalk is the leakage of signals from one channel to another.

MULTIPLEXING

One of the problems of radio or television broadcasting is to keep the transmitted signals from interfering with each other. To keep them from doing so they are each assigned different operating frequencies. You select any of the transmitted channels you want by means of a tuning device on the front panel of your radio or tv set.

Telephone systems use a similar technique, with the difference that the signals are sent over wires as well as the use of microwave radio. The requirement is to have as many voice signals sent over as few wires as possible and to do so in such a way that each voice signal does not interfere with any other transmitted voice signal.

To be able to do so all the voice signals are multiplexed, that is, changed in frequency. Thus, one voice could be raised to 70,000 Hz to 73,000, Hz and another from 73,000 Hz to 76,000 Hz. Some individuals, of course, have a wider speaking frequency range. Thus, one person could have a range of 4000 Hz, another 3500 Hz, and still another only 3000 Hz. When these are multiplexed, that is, changed to higher frequencies though, they will all have the same frequency range or bandwidth. The individual having a wider range speaking voice will have his voice frequencies compressed.

The frequencies used in multiplexing are intended to keep any one voice from interfering with another, and to permit many voices to coexist along the same pair of wires. The multiplexing frequencies are well outside the human hearing range and so are inaudible. And, while voice signals may be multiplexed during transmission, they must be changed back to their original signal frequencies before being delivered to the earpiece of the receiver in the telephone.

When the electrical signal representing your voice is first sent out via telephone it travels via a pair of wires to the Central Station of your local telephone utility. Separate wire pairs are used for each telephone. Multiplexing devices can be used at the Central Station for continuing the travel of your particular electrical signals to other stations. After the multiplexed signals arrive at the Central Station nearest to the home of the person you are calling, they are then changed back to their original form, proceeding through a wire pair to the called telephone.

Fig. 9-12 shows an arrangement for the routing of long distance calls from a business or a home. For homes, individual wires lead from the telephones to the Central Station. Tis is the same for a business except that the lines from the business phones may lead to a private exchange. In some arrangements the phones used in a business also function as exchanges.

From the Central Station the calls are routed to a trunk exchange. The telephone signals are multiplexed and are then transmitted using a microwave radio transmitter. The dishes shown in the drawing are circular reflectors which help aim the signals in a particular direction.

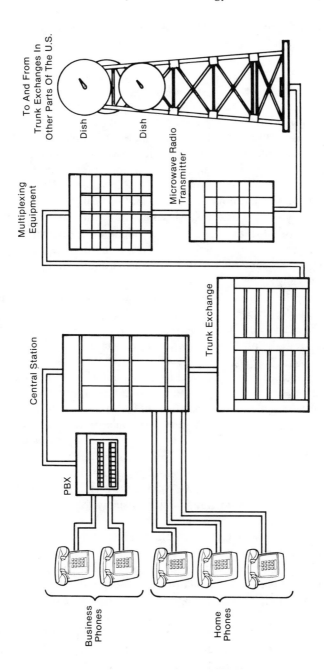

Fig. 9-12. Facsimile transmission system.

Two dishes are shown. One is used for transmitting, the other for signal reception.

FACSIMILE

While videotelephony, a system in which you can see as well as hear persons calling you, is not available, it is possible to transmit pictures over telephone lines using an arrangement known as *facsimile*. These pictures can be photographs, newspapers, or documents containing text material only or text material plus pictures.

One method of facsimile transmission, (also called fax) is shown in Fig. 9-13. The document whose contents are to be transmitted is wrapped around a drum, a cylinder capable of revolving. As the drum turns, it is scanned, line by line, by a photocell (a photoelectric detector) which converts the light reflected from the document into a corresponding electrical signal.

A strong light is made to shine on the document. With this arrangement the maximum amount of light is reflected from white areas, the least from black. Between these extremes, the amount of light reaching the photocell depends on the reflectivity of the scanned surface.

The electrical signal produced by the photocell is rather weak and so it is brought into a preamplifier, a solid-state device that works in somewhat the same manner as an amplifier in audio hi/fi system. The amplified signal is then sent into a modulator for loading onto a radio-frequency carrier. The method of modulation can be either am (amplitude modulation) or fm (frequency modulation) using techniques similar to those of broadcasting stations.

The modulated signal can be transmitted via telephone lines or through the use of radio transmission. Usually the transmission is a combination of both. The facsimile signal, at the output of the modulator, is fed into wired lines (telephone lines) and from those lines to a radio transmitter.

At the receiving end the radio signal is picked up by a receiver which may contain a demodulator or the demodulator may be a separate unit. From the demodulator the facsimile signal is fed into a driver, a signal amplifier. The amplified signal is used to drive a stylus which reproduces the transmitted document, line by line, on a sheet of paper wrapped around a revolving drum. In some units the signal, instead of driving a stylus, is used to reproduce the text and photos optically on photographic film.

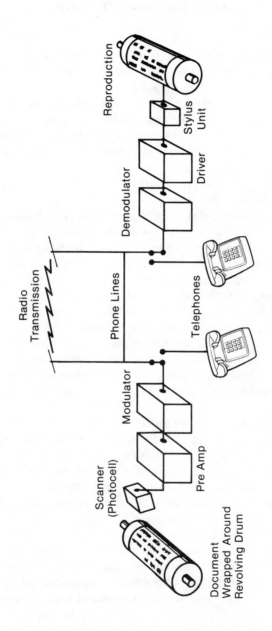

Fig. 9-13. Microwave transmission of multiplexed telephone signals.

TELEPHONY VIA SATELLITE

When the Soviets launched the first manmade satellite Sputnik, in 1957, the opportunity for another long-range method of communications became apparent. In 1958 a U.S. satellite, called *Score*, transmitted a voice recording of the President.

This satellite, and others that were subsequently launched, sometimes called a "bird," was put into a rather low orbit and so it circled the earth at a rapid speed. This interfered with its ability to function as a signal-handling satellite, since the curvature of the earth interfered with signal reception.

To overcome the signal transmitting difficulties of satellites a more effective communications satellite, called "*Syncom*," was put into a geosynchronous orbit, also called a geostationary or fixed orbit. This is an orbit in which an object in space has a rate of speed such that the satellite seems to remain fixed in space. Both the earth and the satellite are not stationary, but their speeds relative to each other are fixed. Actually, because of its larger orbit, the speed of the satellite is somewhat greater than that of the earth's rotation.

In operation, the satellite behaves as though it were a tremendously high antenna. Signals, transmitted to the satellite from an earth station are picked up by the satellite, amplified, and transmitted back to earth. But because of its height, 22,279 statute miles above the equator, the signals from the satellite can be sent to any area of the U.S. Long distance and data transmission can be readily achieved in this way.

In 1965, nineteen countries formed a group called *The Third International Television Satellite Organization* or *Intelsat*. Its purpose was to provide, in addition to television signals, telephone and data communications on an international basis. The number of countries in Intelsat has now grown to over 100, is equipped with 5 satellites and has more than 250 ground stations. These satellites are fifth generation and so are called Number V.

The transmission of data by satellites has a number of advantages. Satellite signals are not affected by the weather, sunspot activity, or the time of day. Because of the very high frequencies that are used (measured in gigahertz) more bandwidth is available. The greater the bandwidth, the larger the number of channels that can be accommodated.

CHAPTER 10

MOBILE TELEPHONE COMMUNICATIONS

A mobile telephone is any telephone that can be used in a vehicle, or which can be carried either by hand or vehicle, and which is not directly connected to a telephone line. The vehicle can be a car, an RV (recreational vehicle), van, truck, boat, plane, or any other kind. A cordless remote handset could be considered mobile since it can be carried from place to place. A walkie talkie is a mobile unit since it is constructed to be portable and used as such.

Mobile communications implies communications while in motion but that is not necessarily the case. An auto can communicate with a fixed position base station whether the car is parked or is moving. And, since motion is involved, the only way to handle the transmission of the signal is via radio waves. After the telephone signal is received from the mobile unit, it can continue along public utility land lines. Alternatively, mobile communications can consist only of radio transmissions back and forth, with no land lines involved.

BANDS AND CHANNELS

A band is a group of radio frequencies intended for a specific communications purpose and assigned for that purpose by the Federal Communications Commission (FCC). Thus, the 1.7 MHz and 49 MHz frequencies have been selected for cordless telephone systems. These are not single frequencies, but are bands. The 1.7-megahertz band extends from 1.690 to 1.770 megahertz and the 49-megahertz band is from 49.830 to 49.890 megahertz.

These bands are subdivided into five channels. The 49-megahertz band has operating frequencies of 49.830; 49.845; 49.860; 49.875; and 49.890 megahertz. Each channel has a total bandwidth of 0.015 megahertz. A channel, then, is a group of frequencies within a band. The number of channels within a band, and their range, are established by the FCC.

MOBILE AND BASE STATIONS

A mobile station is one that is capable of motion, although the mobile unit may or may not be in motion. The base station is a fixed-position setup. Consequently, in mobile telephony we have a number of communications possibilities: from mobile to base, from mobile to mobile, and from base to base. Communications is by radio and/or by telephone land-lines.

THE MOBILE RADIOTELEPHONE

All land-type telephones are duplex. You use them to talk and to listen just as if the person you are communicating with is standing right next to you. It is a two-way conversation, with as many interruptions on either side as wanted.

A mobile radiotelephone is different since it is often simplex (although some have a duplex capability). Simplex means one person talks at a time and does so by depressing a push-to-talk button. Unlike duplex, only one person can talk at a time. Mobile telephones, then, are transceivers, a word that is a contraction of transmitter-receiver. It can work as a transmitter or as a receiver, but usually cannot do both at the same time. Your land-based home or business telephone is not a transceiver, but rather a full transmitter-receiver with both capable of operating at the same time, something many mobile transceivers cannot do.

A mobile radiotelephone is a two-piece unit. One section is a box containing the radio transmitter and receiver, and the other is a handset that looks exactly like the usual telephone handset except it is equipped with a push-to-talk button. In mobile telephony, the handset is sometimes referred to as a control head.

POWER SOURCE

Unlike a land-type telephone which receives its operating power from the telephone lines, the mobile radiotelephone must be supplied with its own source of power. This is generally 12-volts dc and is furnished by a storage battery. Some vehicles have the storage battery with the negative terminal connected to the frame, and is then known as a negative ground. Others have the positive terminal attached to the frame and so are called a positive ground. It is important to know, when buying a mobile radiotelephone, whether it is designed to work with a positive or negative ground. Some equipment can be switched to use either type.

OPERATING MODES

A mobile radiotelephone has two operating modes — receive and transmit. When in the receive mode the transceiver uses a relatively small amount of current, generally in the order of ½ ampere. However, in the transmit mode the amount of current demand on the battery is much more substantial and can be several amperes. While this is a heavy current drain, it is an intermittent one since this load decreases when the push-to-talk button is put in its receive position. However, with batteries that are quite worn or which are trying to meet current demands from other loads, such as lights, the current demand by the transceiver in the transmit mode may not be met and the transmitted signal will be much weaker.

OPERATING FREQUENCIES

The operating frequencies of mobile radiotelephone units are those established by the Federal Communications Commission (FCC). These frequencies form bands extending from 150 MHz to 174 MHz and 450 MHz to 512 MHz. These are rather high frequencies but have the advantage of permitting the use of antennas which are quite small.

With simplex operation of a mobile radiotelephone (sometimes called a radiophone) the same antenna can be used for receiving and transmitting. When the push-to-talk button is depressed, the antenna is switched over from receive to transmit and vice-versa.

The separation frequency between the transmitted and received signals is 5 MHz. Thus, if your transceiver is working on a transmitting frequency of 455 MHz, then the frequency of the received signal could be 460 MHz. These are known as carrier frequencies. Their only function is to carry the voice signal back and forth between the two stations involved in the communication.

MODULATION

The process of loading speech onto a carrier wave is known as modulation. Modulation is limited to a maximum of plus/minus 5 kiloHertz (5 kHz) for radiotelephony. When the modulated radio-frequency carrier is picked up by the receiver section of the transmitter, the carrier is eliminated, while the voice modulation is retained, is amplified and used to operate the speaker in the handset.

PTT VS VOX

PTT is an abbreviation for *push-to-talk*. All simplex type devices are PTT units. A VOX unit is one that is *voice actuated*. Some components used in telephony are equipped with a VOX bypass. With these the component can be switched to manual (PTT) operation. The user would then need to press the push-to-talk control before transmitting. PTT is widely used with fm transceivers, other types of mobile radiotelephones, and with CB units.

SHIPS IN HARBOR

When ships are docked at a pier, land lines are connected and so it is possible to have direct telephone service.

SHIP-TO-SHORE CALLS

Radiotelephone calls can be made to or from ships and smaller pleasure craft when equipped with mobile radiotelephone units. Reduced rates do not apply to ship-to-shore calls. Calls are made through a shore-based Marine Operator. This means all such calls are operator assisted. Communications with the Marine Operator is via radio, but the call is completed through land lines.

MOBILE RADIOTELEPHONE FOR MARINE USE

The component shown in Fig. 10-1 is a marine radio transceiver using 55 channels for transmitting and 75 channels for receiving. Operating from a 13.6-volt dc source it uses a current of 1 ampere on standby and 5 amperes when transmitting. Its maximum output transmitting power is 25 watts.

The unit, however, is more than a transceiver. It has a computer-controlled direction finder utilizing 36 scanning yellow light-emitting diodes. It is a scanning type and can cover 15 vhf channels per second. Additional features include touch entry for all U.S. or international marine channels and four weather channels. A direction finding antenna is supplied.

MOBILE SERVICE

Calls can be made to and from any vehicle: truck, van, taxi, limousine, train, etc., provided these vehicles are mobile radio equipped. Calls are made through a Mobile Service Operator and may include a Long Distance Operator. Reduced rates do not apply to these calls. Calls can also be made to certain trains up to a short time before their departure.

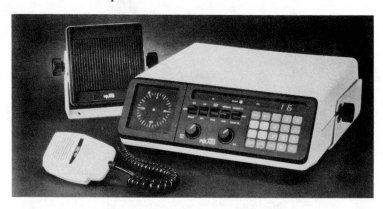

Fig. 10-1. 25-watt marine transceiver. *(Courtesy Regency Electronics, Inc.)*

HANDHELD TRANSCEIVERS

Sometimes called walkie-talkies, a handheld is a combination transmitter/receiver, and is ofen called a transceiver although it is just one type of transceiver. Such units are well suited for communications between persons who are separated by distances of about 1 mile or so, but who do not have convenient access to a land-line telephone. Like a cordless telephone, a transceiver can be in touch with a fixed base station, but unlike cordless systems, all elements of the transceiver arrangement can be mobile. Thus, with the help of a pair of transceivers, two persons can be in communication with each other, although both may be mobile.

Transceivers can be used in countless applications, such as security, warehouse communications, film and stage production, hiking, cycling, boating, paging operations, monitoring children and invalids, and in surveying.

The transceiver shown in Fig. 10-2 is designed for marine use. It can transmit on 54 channels; and receive 75 channels. On standby the

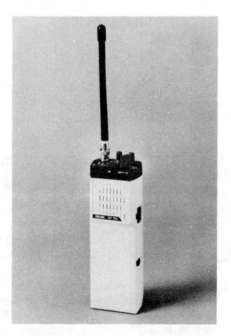

Fig. 10-2. Hand-held transceiver for marine use.
(Courtesy Regency Electronics, Inc.)

unit uses 40 milliamperes of current, and in the transmit mode, 0.9 ampere. Its maximum transmission power is 3 watts.

Since the transceiver does not use telephones, there is no need to supply the local phone utility with an FCC registration number. Further, no FCC license is required. However, before investing in a transceiver, make sure it is FCC approved.

The minimum equipment required for transceiver use is a pair. More than two units can be used, though.

TRANSCEIVER OPERATING FREQUENCIES

Transceivers can operate in the low-power, license-free 49 MHz band. For a representative unit the frequency range could be 49.820 to 49.890 MHz. Frequency modulation (fm) is used.

TRANSCEIVER SPECIFICATIONS

Fig. 10-3 shows another transceiver intended for land use. The operating controls are near the top of the unit. The power source

Fig. 10-3. Handheld transceiver for land use weighs less than 11 oz.
(Courtesy Maxon Electronics, Inc.)

consists of four AA size penlight batteries, with a total of 6 volts dc. When the unit is in its receive mode its current drain is 35 mA (35 milliamperes or 0.035 ampere). During the time it is transmitting the current demand almost doubles and is 60 mA. The battery power supply can deliver up to 8 hours (or more) of operation. A dc charger is included for recharging the nickel-cadmium batteries.

The operating range of this transceiver is about ¼ mile (400 meters) but under optimum conditions can be as much as ½ mile. The whip antenna can be clipped over the headset, or released upright for maximum range. The headset is actually a three-piece component, consisting as it does not only of the headset, but an electret microphone, while also working as a support for the whip antenna. VOX bypass is still another feature, permitting switching to the manual (PTT) mode. In this mode the user presses the push-to-talk switch before transmitting.

ADVANTAGES OF SIMPLEX

Some of the advantages of simplex have been touched on previously. Some of the components and circuits in the transceiver can be made to do double duty. Since reception and transmission do not occur simultaneously, a single frequency channel can be used without the possibility of interference. Simplex is also economical of operating power.

HALF DUPLEX

Telephones used in the home and in business are full duplex, and so communication occurs naturally. However, in mobile applications the telephone must be battery powered and so there is a need for conserving battery power. The maximum amount of power is used when the telephone is in its transmit mode.

For more economical battery power use, a technique called half-duplex is sometimes employed. In this arrangement, as in simplex, the telephones are equipped with a push-to-talk button. There are two communications channels so either party can interrupt at any time. Since a push-to-talk button is used, there is no transmitter power demand until that button is depressed.

MOBILE TELEPHONE LIMITATIONS

Existing mobile radio telephone systems are limited. It is estimated such systems are capable of handling only 21 simultaneous conversations in an area of 50,000 square miles.

To understand the limitations of the present mobile telephone system consider that a telephone company could conceivably have a customer waiting list of more than 1,000 who have been put on a "hold" list for more than three years. And, compared to the cost of in-home telephone service, the price isn't cheap. The system has an installation charge of $195.50 plus a charge of $149.7026016 for basic service. But even with the delay and the high cost, some mobile radiotelephone users have reported problems in obtaining operating frequencies.

MOBILE RADIOTELEPHONE FOR AUTO USE

While a mobile radiotelephone used in automobiles, vans, trucks or other mobile land vehicles may resemble the telephone used in the home, there are quite a few differences. Thus, the unit illustrated in Fig. 10-4 has an illuminated key pad, useful since the interior of the vehicle is not ordinarily illuminated while driving. Still another driver-oriented feature is that all operating controls are right on the handset. This means that all calls can be placed or received using one hand only. A digital display shows the number called, in addition to "access," "busy," "incoming call," and "channel selected." With the addition of a remote speaker it can place calls while on-hook. You can also program up to 10 telephone numbers of up to 14 digits in length into memory for recall at a later time.

With a land-based wired telephone, such as the type you use in your home, not being able to obtain a dial tone is an extremely rare occurrence. The presence of a dial tone indicates you have a clear channel to the telephone Central Office and that you can proceed with your telephone call. In effect, what you have is an unlimited number of channels.

With mobile radiotelephony the number of available channels is limited. This means that when using the mobile radiotelephone there will be times when you will need to wait for a clear channel. The unit shown in Fig. 10-4 has a feature known as *call queuing*. If a busy tone is heard when placing a call due to all channels being occupied,

Fig. 10-4. Mobile radiotelephone for vehicle use. *(Courtesy Harris Corp.)*

you can depress a pair of selected buttons sequentially and, when available, a channel will be secured, as indicated by receipt of a dial tone.

To prevent theft of service this component has an electronic lock, using three digits. The user must enter a 3-digit access code before calls can be placed. However, calls can still be received with the unit in its locked position.

This component covers vhf operation in the vhf range of 150 to 174 megahertz and the uhf range of 450 to 512 megahertz. It can be mounted in the car trunk, under a seat, or any other areas of the vehicle. Lock-in mounting secures the radio to its mounting tray, allowing it to be removed only with a key.

Fig. 10-5 illustrates how communication is achieved between autos or between an auto and a base station. In both instances two different operating frequencies are used: one for transmitting the signal, and the other for receiving it. Consider, first, car to car communications. The calling party, shown by the car at the left, transmits on a high frequency. The called party receives this high-frequency signal but answers by transmitting on a lower frequency. Because two different frequencies are used, the operation is duplex. However, it does

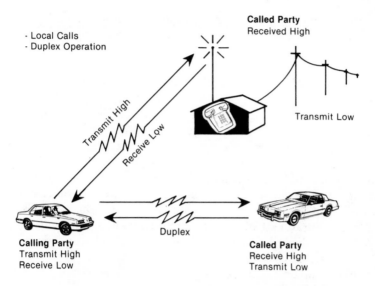

Fig. 10-5. Vehicle telephone duplex transmission method.

require two channels. Further, it means that both channels must be "open," that is, not in use by any other pair of vehicles. And, as long as the two vehicles shown in the drawing are talking to each other, those channels are "tied up," not available for any other use. Because of this, it is urgent that talking time between the two cars be kept as brief as possible. It also means that those in the two cars who want to talk to each other must first search the two channels to make sure they are not occupied. If just one channel is available then duplex communication is not possible.

The same illustration shows how a vehicle can communicate with a fixed-position base station. Again, a pair of channels are needed for duplex and once again a search must be made to ensure a pair of clear channels.

There are a large number of channels available for mobile radiotelephony, and, as indicated earlier, the component shown in Fig. 10-4 has a 128 channel capability. For this reason a scanning feature (called queuing) is desirable. The component automaically searches up and down the frequency band for a free channel and will seize the first one that becomes available. The alternative is to scan the mobile bands manually, a tedious process.

The advantage of using a base station is that a telephone call from a vehicle can continue along land telephone lines, thus giving the mobile operator an opportunity to call his home or office. In this way a car can call any land-based telephone anywhere in the world.

LIMITS OF MOBILE COMMUNICATIONS

Two vehicles can communicate with each other via radiotelephone assuming they are within range. This means both mobile setups should have approximately equal transmitting strengths and that the receivers of both are equally sensitive. Communications will be limited to the receiver sensitivity and transmitting strength of the weaker unit. Even with these factors there is also the problem of terrain. It is possible for the signals to be blocked by a building or hills. The signals, as transmitted, have a certain amount of directionality, depending on how the transmitting antenna of each car is positioned. The signals are also subject to electrical interference. To ensure better mobile communications for land vehicles various systems have been proposed. One of these is cellular radio — the other Personal Radio Communications Service.

CELLULAR RADIO

With a cellular system which has been proposed, communities would be divided into districts or cells, hence the name. Each district or cell would have its own tower for the pickup and transmission of signals. National communications could be had with a cellular system through connections to a long distance microwave network.

Thus, cellular radio will provide people on the move with more effective mobile telephone service. A call will travel over local telephone lines and will then be sent to the designated mobile unit by the nearest transmitter. Mobile units moving from one area to another will be tracked and the call will be switched without interruption.

Cellular mobile radio service is directly dependent on the FCC. That government agency has authorized cellular service and it has done so by setting aside enough radio operating frequencies for two different systems in each city, nationwide. Those freqencies that have been made available will be regarded as a franchise, but one which may not be subdivided.

Fig. 10-6 details the cellular concept. Visualize a large geographical area divided into groups of hexagonal cells adjacent to each other. At

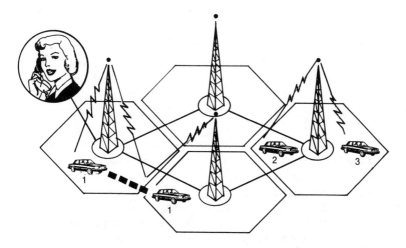

Fig. 10-6. Operation of a cellular telephone system.

the approximate center of each cell there will be a radio transmitter and a computer controller. Each transmitter would be connected to any other transmitter by telephone lines.

As an example, locate car No. 1 in the first hexagon at the left. The occupant of that car is talking to a caller shown in the circled inset. As car No. 1 moves out of the first hexagonal area into the next, its transmissions are picked up by the centrally located transmitting tower in that hexagon. Using computers, the operating frequency originally used is now shifted, but done so rapidly and smoothly that the parties engaged in the telephone conversation never notice it. The conversation is not interrupted in any way.

Because the transmitting power of each central tower is very low there is no interference from one area to the next. A large number of frequencies are used, but each of these is on a local basis. In effect, the method is a relay system with the signal being handed along from one area to the next. The user of a mobile unit will be able to talk locally from his vehicle, or long distance.

While car No. 1 in Fig. 10-6 has moved from one tower serviced area to the next, car No. 2 has traveled much less and remains within the area of its tower. Car No. 1 could also make contact with car No. 3.

As indicated earlier, cellular radio is considered a solution to the present problems of mobile radio, an expensive, difficult-to-obtain service.

RCCs

An *RCC* is a radio common carrier consisting of companies that manufacture mobile radio equipment and operate paging systems. With a cellular radio setup, two factors will be involved. One of these will be an RCC, and the other a local telephone utility. Ultimately, there will be a complete network of cellular arrangements across the nation.

Thus, if you make a telephone call using the mobile radio telephone system in your car, that call will be picked up by the receiving antenna of the local telephone company. It could then possibly be transmitted by land wire or radio to some distant city. There it would be rebroadcast by the transmitting antenna of a centrally located local telephone company. The telephone call would then be received by some mobile radio receiver, located in a car, boat, a truck, or a van. Communications could thus be established between widely separated mobile units.

Note the difference between cellular radio and the existing mobile radio setup. Presently, we have one, large, high-powered centrally located transmitter. But even with high-power transmitters, signal strength falls off rapidly. With cellular radio cells will be established in various areas. These will be low powered. Telephone calls picked up from mobile units will be transferred, by a computer system, from cell to cell. And, since the power of the centrally located transmitter will be so low, the same operating frequency will be usable for other services in the same metropolitan area.

PERSONAL RADIO COMMUNICATIONS SERVICE

Nonmobile radiotelephony is such an established fact we literally take it for granted. We assume communications via telephone is practical, feasible, and possible. But communications from a moving vehicle or some other vehicle or between a vehicle and an in-home or in-business telephone is not all that simple. The only way a telephone can communicate from a moving vehicle is via radio, but the number of frequencies available for such communications is severely limited.

CB radio is a case in point. Here a number of frequencies have been established for what is usually auto-to-auto communications, but anyone who has ever used the CB bands is aware of the difficulty of getting a clear channel, and of keeping that clear channel long enough to complete communications and avoiding interfering signals. Further, these channels have a limited practical operating range and while there are examples of illegal long distance communications, these are unreliable and sporadic, certainly not dependable.

One method, proposed by GE, called the *Personal Radio Communications Service (PRCS),* is designed to augment the limited communications capabilities already available to consumers for road to home service. It provides the user with an affordable, quality service for private communications within a person's normal driving area.

People frequently must communicate while on the road so there is a need for an affordable private mobile communications system that will augment CB, general mobile radio services, and cellular mobile telephone systems.

Sixty million American households use their vehicles for personal reasons very day and thirty-five million use vehicles to commute. Forty percent said they need communications capabilities while in their vehicles. There are presently only two basic types of in-vehicle communications systems available to the general public. One is mobile telephones, expensive to buy, plus an equally expensive monthly service charge. The other system is CB radio, with the limitations mentioned previously.

The anticipated price for a PRCS system including a vehicle unit and a base unit will be around $400. With this system you will be able to communicate from your vehicle to your home, office or other base location. You will be able to telephone from your base location to your vehicle. You will be able to communicate from one vehicle to another, if both are PRCS equipped. And anyone who has a telephone will be able to reach you in your car when you are within range.

The basic PRCS system will give you a range of about 3 to 5 miles from your base station. The PRCS System also provides for the development of local repeater services throughout the country. As these become available, PRCS owners will be able to extend the range of their system to about 15 miles by subscribing to a repeater service.

Fig. 10-7 shows a photo of a PRCS base station. To be installed in home, the base station, in conjunction with a vehicle mobile unit, will provide private, addressable mobile communications. Fig. 10-8 is a picture of a PRCS mobile unit, to be installed in motor vehicles.

Depending on how PRCS will finally develop, it may be operated

Fig. 10-7. Personal Radio Communications Service base station for installation in the home. *(Courtesy General Electric)*

Fig. 10-8. Mobile unit to be installed in motor vehicles for proposed Personal Radio Communications Service. *(Courtesy General Electric)*

simplex or duplex. The drawing shown earlier in Fig. 10-5 is an example of duplex operation. Two frequencies are required: one for transmission, the other for reception. With this setup two automobiles could have communications. The calling party could speak directly to the called vehicle. For longer range transmission, the vehicle, such as the one shown at the left, could call a base station. The base station could represent the terminal point of the call, or it could be used to forward the call to some other in-home or in-office telephone.

Fig. 10-9 shows a two-frequency simplex arrangement. In this ar-

- Repeater
- Two Frequency Simplex
- Calling Party Must Switch Bands To Accomodate System

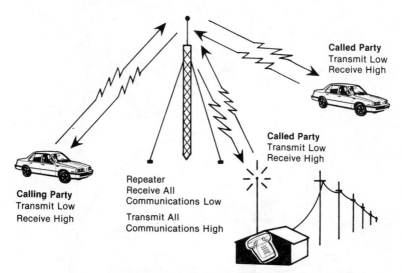

Fig. 10-9. Vehicle telephone simplex transmission method.

rangement the operator of the mobile communications system in the car puts through a call to a repeater station since the range between the two cars is too great for direct calling. The repeater station can send the mobile telephone call on to the called party in another car or can send the message along to a base station for further routing.

Still another communications possibility is shown in the drawing of Fig. 10-10. For communications between two vehicles a working distance of about 2 miles is a reasonable expectation. Actual working distances would depend on the terrain, electrical noise levels, and operating conditions of the working units. The strength of a signal transmitted by the mobile unit can vary from one area to the next. If the vehicles operate near the outer limits of their transmission and reception ranges, communications can become erratic, or the noise level may become too high.

The working distance, the distance in miles over which effective communications can be established between a mobile unit and a base station, is about 5 miles. Between a vehicle and a repeater the range is about 15 miles. Using telephone company long distance lines a telephone call could be made from any vehicle or to any land telephone in the U.S. or internationally.

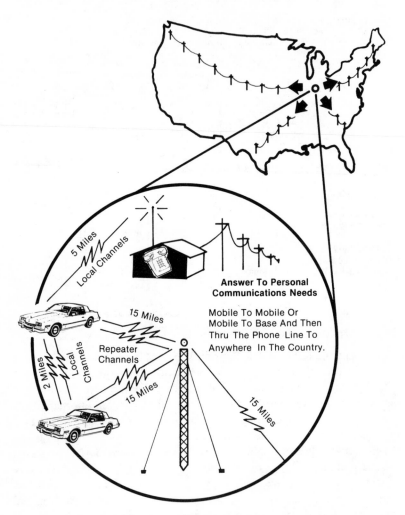

Fig. 10-10. Practical communications limits for vehicles.

The advantage of using a repeater station is evident. Assuming an effective range for the repeater of 15 miles, then a pair of mobile radiotelephone equipped cars making use of repeater service could be separated by as much as 30 miles. If we accept 2 miles as the limit of communications without the repeater, the use of the repeater multiplies the communications range by a factor of 15. Also, with the repeater, the signal-to-noise ratio increases meaning that not only is range extended but the quality of the telephone signal as well.

PAGING SYSTEMS

Communication is no great problem as long as you and the person (or persons) you want to talk to are within reach of a landline telephone. But outside the home or office, and not within reach of a public telephone, there is no way without some electronic assistance in which we can know whether someone is trying to reach us.

Policemen in the field and salesmen on the road often use telephones to make contact with a central office, but they are the ones initiating the calls and the calls they make are to a fixed position telephone. A central office trying to reach a policeman or a business trying to contact a traveling salesman are faced with the problem of reaching a mobile person.

A method of doing so is through the use of a radio pager. A radio pager, sometimes called a "beeper," is about the size of a pack of cigarettes. It is actually a small radio receiver that is fixed tuned to a particular frequency. Someone carrying a beeper, a physician for example, can be reached by a signal transmitted by a central station. The central station radiates the signal at the request of a caller. Upon hearing the beep the user of the pager telephones the central station and receives the name and telephone number of the person trying to call him.

Beepers produce a "beeping" sound, hence their name. But some of them can deliver an oral message, although quite short, while some can display a 10-digit number, using liquid crystal diodes, somewhat in the manner of an electronic calculator. The digits can be the message, with the numbers representing a telephone to be called or a code of some kind. Some beepers are alphanumeric, that is, display the letters of the alphabet plus digits, while some are digital only.

The effectiveness of a beeper arrangement depends on a number of factors. The user of the beeper must be within the range of the station transmitting the call signal. The usual range is about 30 miles. This is a form of strictly local communications, but it is in the process of change. There are now some 28 channels being used for this service, with each channel capable of accommodating between 60,000 and 100,000 beeper equipped persons.

The problem with the present pager is that the user needs to telephone a central station, the station that broadcast the signal, to learn the telephone number of the caller. The newer units still beep, although some blink silently, but they display or print out the number for the subscriber to call.

An advantage of the printout beeper is that it may be able to have several call-back number possibilities, eliminating the need for calling the transmitting station and writing these. When the beeper has an alphanumeric display it eliminates the need for the user to make a telephone call to a central station. Thus, a subscriber may get personal messages, corporate memos, stock quotations, sports results, and also the telephone number of a person to call.

Like anything else, beepers have advantages and disadvantages. For the person who is moving about it means it means he or she can be reached by the beeper service. Beepers can also be used to establish contact between a pair of persons who are beeper equipped. But the person carrying the beeper must always remain within radio reach of the transmitting station. The beeper must always remain turned on. And while beepers can be clipped to a belt or carried in a purse, they do occupy space and they do have weight. Also, the sound of a beeper does attract attention. The alternative to the expense of a beeper and the nuisance of carrying it is for the mobile person to use a telephone system, public or otherwise, and telephone home or office. This is not always practical or possible.

DUPLEX WIRELESS INTERCOM

Not all mobile communications units are used in vehicles. It is possible to have a mobile unit in the home through the use of a wireless intercom. The unit looks like a regular telephone, consisting of a handset and a main body. The component does not require any in-home premises wiring. All you do is plug it into the nearest convenient ac power outlet. There is one requirement and that is the telephones must be employed within the ac lines supplied by the same transformer.

The telephone can be carried anywhere in the home or office but can communicate only with similar telephones. More than a pair of such telephones can be used if wanted. And, since telephone company lines are not used there is no need to inform your local telephone utility.

Each telephone unit contains a very-moderate-power fm transmitter with a power consumption of about 2.5 watts, far less than the average night light. However, instead of loading the voice modulated signal onto an antenna, the ac power lines are used instead. The system is duplex and so both parties can use the unit just as they would with an ordinary telephone.

The unit does not have a dialing pad. Instead, lifting the handset from the main body produces a tremolo tone that alerts the called party a call is waiting. The call continues for a period of two seconds. If the called party does not answer, all you need to do is to press down on the hookswitch and then release it. At this time the other party's phone will ring for another two seconds.

GLOSSARY

This is a listing of acronyms, abbreviations, and definitions of words and phrases used in telephony.

A

ac—Alternating current.

access code—Number or a group of numbers that let you connect to a line or service.

accessory—Various devices such as wire, plugs and jacks.

acoustic feedback—A condition in which sound is fed back from the transmitter to the receiver in a telephone, resulting in howling.

active device—A component or circuit that requires a source of power.

add on—A component connected to a phone or phone system for supplying additional features.

add-on dialer—Dialing accessory for converting rotary dial phones to pushbutton.

advanced telephone—A phone having a number of unusual features and generally made in a one-piece style.

am—Amplitude modulation. A type of modulation in which the amplitude or strength of a carrier wave is changed in step with an impressed audio voltage.

analog signal—A voltage that changes in step with the sound producing it.

answer only phone—An extension phone not equipped with a dialing pad. A phone that does not contain a ringer.

audio feedback—A condition caused by putting the receiver of a telephone too close to the transmitter, resulting in howling. This can only be done with two-piece phones.

auto disconnect—A telephone feature that will disconnect after a selected number of rings to a phone that does not answer.

automatic dialing—Memory assisted dialing requiring touching a single pushbutton for making a phone call.

automatic shutoff—A circuit or device for turning off a telephone answering machine at the end of message time and at the end of the tape.

automatic stop—Some answering machines will recognize that a caller has hung up and will stop recording.

B

backspace erase—Used for making corrections when dialing, eliminating need for redialing the complete number.

bandwidth—The frequency range of a group of signals ranging from the lowest to the highest frequency.

base station—A telephone or phone system in a fixed position, in a home or office, for receiving phone calls or for receiving and retransmitting them.

battery-low indicator—An indicator that glows when battery voltage drops to a level that is too low.

baud—The bit per second when all pulses have the same amplitude.

beeper—A compact pocket coder. A device for communicating with a telephone answering machine or a paging service.

binary system—A numbering system using two symbols, o and 1.

bit—Acronym for binary digit.

break position—A switch that is open.

budget phone—Telephone selling for a very low price.

bug—Any device used for surreptitious listening to a telephone conversation.

building service—A service supplied by local telephone companies for home or office wiring and phone layout problems.

bundling—Grouping of charges on a telephone bill.

buzzer—Alarm used on remote handset of a cordless system.

C

cable—A group of two or more wires often surrounded by a protective covering.

call button—Control on a base station used for operating a call signal at the remote handset.

call forwarding—The transfer of incoming calls to any telephone number.

call monitoring—Listening to or recording a telephone call.

call screening—Telephone equipped with device for eliminating unwanted calls.

call waiting—A service that lets calls get through to you even when your line is busy. A special tone to alert you that another call is waiting.

calling card—Credit card issued by a telephone utility.

cancel—A phone feature that stops any operation in progress, eliminating the need for hanging up the phone and waiting for a dial tone.

capacitor—A component or part for storing an electrical charge.

carbon microphone—Transducer which uses a carbon capsule in conjunction with a flexible diaphragm.

carrier—A high frequency wave used to carry much lower frequency audio signals.

carrier current antenna—House wiring used as an antenna.

cassette—Plastic shell housing magnetic tape wound on reels.

cellular radio—A proposed system of mobile communications for automobiles using base stations or retransmission towers.

central office connection charge—Charge for connecting phone lines to a central office switching system.

channel—A single frequency or a band of frequencies. A section of a band of frequencies.

closed circuit—A device having minimum or zero impedance.

code—A series of numbers used for a specific purpose such as access, line selection or phone security.

COM—Company supplying long distance calling services.

compatibility—The ability of two or more components to interface or to work together or which can be interconnected.

component phone—A telephone to which one or more components have been added. A telephone joined with some other device such as an answering machine.

conference calling—An option permitting you to add another party to your telephone call. A method of having as many as 10 telephones interconnected.

console PBX—Desktop switching service.

cord—A length of wire covered by an insulating material.

cord coupler—Accessory for connecting a pair of modular cords.

cord-free phone—See cordless telephone.

corded phone—Phone connected directly to phone lines by pairs of wires.

cordless handset telephone—A cordless phone not equipped with a dialing pad. Comparable to an extension phone, also without a dialing pad.

cordless telephone—A phone that can transmit to a base station or receive phone calls from it. A telephone system consisting of two units: a base or fixed position telephone and a remote or portable hand-held phone.

country code—The access code of a country.

cradle—Part of a telephone used for holding the receiver, or, in some cases, the entire phone.

crossbar switching—Device for connecting pairs of telephone wires in a Central Office.

custom calling services—Various services offered by a local telephone company to aid in making calls.

D

data service—Method of using a pushbutton telephone linked with a computer.

dc—Direct current.

decimal system—A numbering system using 10 symbols from 0 to 9 inclusive.

decoding—Opposite of encoding. Method for changing encoded signal back to original form.

decorator phone—A phone having a distinctive or unusual appearance or made of unusual materials.

dedicated telephone line—A line used for a single specific purpose, such as an emergency.

demodulation—Process of removing an audio signal from its high frequency carrier.

desktop phone—Two-piece phone used on any flat surface, generally a desk or table.

dial tone—An audio current signal supplied by a Central Office indicating the connection of a telephone to a switching system. An indication of an available telephone channel.

digital call counter—A counter on a telephone answering machine that displays numerically the number of calls received.

digital display—Phone equipped with miniature screen for showing numbers. Also used on telephone answering machines.

digital telephony—Transmission of data using binary number system.

direct dialing—Dialing without operator assistance.

double cassette TAD—Telephone answering machine using two cassettes: one for announcements, the other for recording and playback of incoming phone messages.

dtmf—Dual-tone multifrequency dialing.

dual dial phone—Phone having two key pads, side by side.

dual-tone multifrequency dialing—A dialing method using pairs of tones.

duplex—Simultaneous communications in two directions.

duplex jack—Extension cord with modular plug at one end and twin modular jack on the other.

dynamic transducer—Device used as a speaker or transmitter, consisting of diaphragm moving a coil mounted between the poles of a permanent magnet.

E

electronic larynx—A device for people who have lost the use of their vocal cords, enabling them to resume voice communications.

electronic tone ringer—A device for replacing usual phone bell ringer, supplying a pleasant, warbling sound.

encoding—Techniques for changing a voice signal either for purposes of security or for changing an analog signal into digital form.

evening rate—Rate applicable to calls made from 5 pm to 11 pm weekdays Monday through Friday and also Sunday.

extended area service—Payment of a flat fee for all calls within an area larger than that of unlimited local call service.

extension telephone—An auxiliary telephone. It may or may not have a keyboard. If it does not have a keyboard, it is also known as a receive-only phone.

F

facsimile—A method of transmitting pictures and text material via telephone lines.

FCC—Federal Communications Commission.

flaking—Magnetic particles falling away from tape.

flip phone—Small compact phone that flies open when picked up.

fm—Frequency modulation. A type of modulation in which the frequency of the carrier is varied at an audio rate.

FTS—Federal Telecommunications System.

H

hands-off telephone—A phone that does not require the lifting of the handset.

handset—A hand-held portable phone used in conjunction with a fixed position base station.

hang telephone—Phone that is suspended by a hook.

harmonic—The multiple of a frequency.

hold control—Control of a phone for keeping calls on the line until the called party is ready.

hold key indicator—A circuit for avoiding the loss of a call when switching between the phone and intercom functions.

holster—A type of cradle for holding or supporting a one-piece telephone.

hookswitch—A switch for turning the dial tone on or off. Works automatically by lifting the telephone receiver.

Hz—Unit combining frequency and time. Abbreviation for hertz (cycles per second).

I

ICM—incoming message tape.

impedance—The total opposition, measured in ohms, to the flow of a varying current through a conductor, such as a wire, or a circuit.

inside wire charge—Billing for connecting phones to exchange access line.

integrated phone—A telephone containing a number of features generally supplied by an add-on unit.

intercom—In-house or in-office communications system not connected to phone lines.

interdigit time—Time between a group of pulses.

interfacing—Connecting phones or phone equipment to telephone company lines.

international access code—A code consisting of two or three numbers used for initiating a long distance call.

interoffice telephone—A phone not equipped with access to outside telephone lines and used only for communications between offices.

J

jack—A receptacle for receiving a phone plug.

K

kHz—Kilohertz (thousand cycles per second).

L

laser—An acronym of light amplification by stimulated emission of radiation.

last number redial—A phone feature in which the last number dialed is repeatedly recalled until it responds.

led—Light-emitting diode used as an indicator of some function or service.

led reference readout—A light-emitting diode display that shows the number being called.

light ringer—A light substituted for the ringer bell in a phone.

limited voice actuation—Used in connection with telephone answering machines. When caller stops talking the machine automatically stops recording.

line—Pairs of connecting wires from a phone to a Central Office. Also called a loop.

line disconnect phone—Phone equipped wth a line disconnect switch.

line interruption—Asking the operator to interrupt a busy line.

lms—Local measured service.

local measured service—A local call quota system with charge on a per minute basis after quota of calls has been reached.

log—A record of telephone calls.

loop—Pairs of connecting wires. Also known as a telephone line or simply a line.

low use residence plan—A restriction on the number of calls that can be made without additional charge.

M

make position—A switch that is closed.

maximum rate—Full day rate.

MCI—Company supplying long distance calling services.

memo record—Use of a telephone answering machine as an electronic memo reminder.

memory—Circuit in a phone having the capability of storing one or more phone numbers.

memory capacity—The ability of a phone to store a series of telephone numbers. Also refers to the number of digits of a phone number that can be stored and dialed automatically.

memory stringing—Ability of phone to operate its memory bank sequentially. Memories functioning one after the other.

message received alert—Flashing light on a telephone answering machine to indicate a received message.

MHz—Megahertz, or million cycles per second.

mic—Microphone. Also abbreviated as mike. A transducer for changing voice to electrical energy.

microcassette—A cassette which is smaller than a standard cassette. Used in a telephone answering machine.

microphone—Device for changing sound energy to a corresponding voltage.

microsecond—Millionth of a second.

microwaves—Ultrahigh-frequency radio waves.

millisecond—Thousandth of a second.

mnemonic—Any memory aid. A way of helping you remember numbers or facts.

mobile station—A communications system carried in a vehicle or by hand.

modem—A modulator-demodulator. Often combined as a single unit.

modern phone—A phone having a number of unusual features and often made in a one-piece style.

modular adapter—Standard four-prong phone plug with a recess to accommodate a modular plug.

modular coil extension cord—Modular wire with snap-in and snap-out plugs on both ends.

modular jack—A jack designed for receiving a modular plug.

modular jack adapter—Modular wire with a modular plug at one end and a standard jack on the other.

modular telephone wire—Wire with a modular plug on one end and four color-coded spade lugs on the other.

modular wallplate—A decorative rectangular plate for use in connection with a modular jack.

modulation—A process of loading or impressing a voice signal (wave) on a much higher frequency carrier wave.

multifunction phone—An integrated phone having many features such as answering machine, automatic dialer, etc.

multiplexing—The use of different modulating frequencies for the simultaneous transmission of signals.

music on hold—Music supplied via phone to the calling party.

mute control—Control on a phone, or available as an accessory, for preventing phone callers from overhearing conversations in the room you are calling from.

N

Network One—Company supplying long distance calling services.

night rate—Lowest of all rates, effective Monday to Sunday, from 11 pm to 8 am.

nonlisted semiprivate number—Nonlisting of your phone number in the phone directory, but with that number still available through Directory Service Assistance.

O

O—Used on a telephone pad for dialing the operator or as the digit zero.

OGM—Outgoing message tape.

ohm—Basic unit of resistance.

on-hook dialing—Dialing with the telephone on hook.

on-hook function key—Used on cordless units that are hook mounted. By depressing this key you can dial without lifting the remote unit.

on-location bug—A telephone listening device in the same room with the phone or in an adjacent room.

one-piece telephone—Telephone with the receiver and transmitter in a single housing.

open circuit—A device having an infinite amount of impedance.

operator assisted calls—Calls requiring operator assistance. These carry the highest rates.

optical fiber—A light-carrying glass thread using laser beams for communications.

oscillator—An electronic generator of ac waves.

outpulse dial phone—Phone equipped with pushbutton key pad but producing the equivalent of rotary dial pulses.

P

PABX—Private automatic branch exchange.

paging system—System using portable beepers and a centrally located station.

parasitic tap—An eavesdropping device that uses telephone line voltage as its source of operating power.

party line—Sharing phone line services with another family. Sometimes called a two-party line.

passive device—A component or circuit not requiring a source of power.

PAX—Private automatic exchange.

PBX—Public branch exchange.

personal access code—A code, consisting of one or more digits, to prevent receiving unauthorized calls.

personal radio communications service—Proposed system of mobile communications for cars. System makes use of centrally located repeater stations.

phone pad—The telephone keyboard.

PLE—Public local exchange.

portable tone dialer—A tone dialer that can be carried and used on any phone.

power indicator lights—Lights that glow when power is supplied to a unit from the ac lines.

pps—Pulses per second.

PRCS—Personal radio communications service.

premises wiring—Conductors (wires) in a home or office for connecting telephones to phone company lines.

pressure sensitive pad—A phone pad that needs only a light finger touch to be operative.

private line service—Used for the transmission of calls and data with no time limit.

private unlisted phone number—A phone number not listed in the telephone directory nor available through directory assistance.

programming—The process of putting selected phone numbers into the memory of a telephone.

ptt—Push-to-talk control on a simplex phone system.

pulse—A momentary flow of current characterized by a sharp rise and fall.

pulse frequency—Number of pulses per second.

pulse selector switch—Control that selects either 10 or 20 pulses per second.

push-to-talk switch—Abbreviated as PTT, and used for simplex operation.

Q

quick touch call answer—In hands-off telephone touching any numerical button for answering.

R

range alarm—Cordless phone equipped with warning tone that sounds automatically at the limit of communications range.

range extender—Supplementary antenna for cordless telephone system.

RCC—Radio common carrier.

rebuilt phone—A used telephone that may have been repaired.

receive only telephone—A phone that does not contain a ringing circuit. A phone that cannot be used for making outside phone calls.

recharging indicator—An indicator that glows when a remote handset is positioned in its cradle, indicating that charging is taking place.

redial key—Key on a cordless phone that controls one number in the memory.

remote bug—Bug operated at some distance from the telephone.

remote handset—See handset.

remote station—A station located at some distance from a base station. Remote station is a telephone setup that is often mobile.

repeater station—Facility equipped with solid-state amplifiers for strengthening and retransmitting telephone signals.

repertory dialing—See memory capacity.

resistance—Opposition to the flow of an electric current. Basic unit is the ohm.

rf—Radio frequency. Usually refers to a high-frequency wave.

rf sensitivity switch—Switch for controlling the sensitivity of a receiver.

ring control—Control on a telephone answering machine that permits adjustment of number of rings before machine starts functioning.

ringer—Device in a phone to alert the subscriber of the presence of an incoming call.

ringing generator—Device in the Central Office for producing a ringing signal in a phone.

ringless phone—Phone equipped with a switch for turning the ringer off.

rotary dial—A telephone dial which is turned, producing pulses.

routing code—The code used in a foreign country for sending a call to a local exchange.

S

scrambler—Device for altering a voice signal to deter surreptitious listening to a phone coversation.

secure override—A wireless telephone system feature which automatically releases a phone for the user of a second handset.

selective erase—A feature of some telephone answering machines that permits erasure of any recorded message anywhere on the tape.

service order charge—Service charge made by phone company for installation or changes in phone service.

ship-to-shore call—A phone call from a vessel to a land-based phone center.

sidetone—Current sent from the transmitter in a phone to the receiver to enable the user to speak at the same voice level as used without the telephone.

silencer cord—Phone cord equipped with a switch for turning phone ringer on or off.

simplex—A type of communications in which only one party can talk at a time.

single cassette TAD—A telephone answering machine using a single tape for announcement and recording of incoming messages.

single-way recording—A telephone answering machine that can record incoming messages only.

speed calling—Calling a number by depressing just 1 or 2 keys. A way of dialing phone numbers, including emergency numbers, that let you reach a called number rapidly.

SPRINT—Company supplying long distance calling services.

squeeze tab—Tab on a modular plug for holding plug securely in position.

standard cassette—A cassette used for recording and playback in audio cassette decks and in some telephone answering machines.

standard jack—Receptacle with four holes for the insertion of a standard plug.

standard telephone plug—Four-prong plug with standard three or four wires.

step-by-step switching—Device for connecting pairs of telephone wires in a Central Office.

suspended rates—Telephone charges for temporary discontinuance of phone service.

switchable dialing—Dialing that can be changed from rotary to touch tone.

switchboard—Device for connecting telephone calls.

switching—Making a connection. A system of connecting pairs of telephone lines.

switching time—Make and break times of a switch.

T

tad—Telephone answering device.

tap—A physical connection to a phone line for unauthorized listening to phone conversations.

tape recorder detector—Device for determining the presence of a tape recorder associated with a bug.

tdd—Telecommunications device for the deaf.

telecommunications device for the deaf—Special equipment installed by the phone company for the hard of hearing.

telephone amplifier—Accessory for adding amplifier to phone.

telephone answering machine—Device equipped with one or more cassettes for answering telephone calls and recording messages.

telephone handling charge—Charge made for each phone the telephone company connects.

telephone line—See line.

telescoping antenna—An antenna consisting of several sections which can be extended or retracted.

teleterminal—Combined computer terminal and telephone.

TELTEC—Company supplying a long distance calling service.

three-way calling—With this service you can add a third party, or more, to a phone conversation.

time pause control—Phone feature that automatically allows the correct time lapse between an access code and the phone number.

timer—A phone feature that times all outgoing calls automatically.

toll restrictors—An arrangement with the phone company to permit some long distance calls to be made, but not others.

tone digit assignment—Use of low and high frequency tones for dual tone multifrequency dialing.

touch dialing keys—Key pad on front of a remote handset.

touch sensitive dialing—A type of light-touch dialing.

transceiver—A combined transmitter/receiver.

transducer—Component that changes one form of energy to another.

trickle charge—A technique in which the battery being charged receives a small, continuous charge. A method of "floating" a battery across a charger.

two-party line—See party line.

two-piece telephone—Telephone in which the receiver and transmitter are housed separately. Also a phone using a single piece housing for receiver and transmitter but using a separate base for the hookswitch.

two-way recording—Feature of a telephone answering machine that permits recording incoming and outgoing phone conversations.

U

uhf—Ultrahigh frequency.

universal combination jack—A jack that will simultaneously accommodate a standard and a modular plug.

unlimited local call service—Payment of a flat fee for all calls within an area defined by the local telephone utility.

V

varistor—A type of variable resistor used in telephones to compensate for the resistance of different lengths of phone lines.

vhf—Very high frequency.

video telephony—A telephone system in which both voices and views of the persons using the phone are transmitted and received.

voice operation—Hands-free operation.

volt—Basic unit of electrical pressure.

vox—Voice actuated.

W

walkie-talkie—A mobile communications device having a limited communications range and not using telephone lines.

wall bug—Bug that can be attached to a wall by a suction cup.

WATS—Wide area telecommunications service.

wireless telephone—See cordless telephone.

WU—Company supplying long distance calling services.

APPENDIX

MANUFACTURERS AND SUPPLIERS

A

All Channel Products
Div. Electro Audio Dynamics, Inc.
Bayside, NY 11361

Almicro Electronics, Inc.
725 Wall St.
Winnipeg, Man., Canada R3G 2T6

American Telecommunications Corp.
9620 Flair Drive
El Monte, CA 91731

American Telephone &
Telegraph Co.
Consumer Products Div.
5 Wood Hollow Rd.
Parsippany, NJ 07054

Anova Electronics,
Div. Dart & Kraft, Inc.
Three Waters Park Drive
San Mateo, CA 94403

APF Electronics, Inc.
1501 Broadway
New York, NY 10010

Arista Enterprises, Inc.
125 Commerce Drive
Hauppauge, NY 11788

Arrow Trading Co., Inc.
1115 Broadway
New York, NY 10010

Assurance Industries Co., Inc.
5353 Almaden Exp. E-47
San Jose, CA 95118

Autumn Company,
Oak Park, MN 48237

T.A.D. Avanti, Inc.
(See Record-A-Call)

B

Buscom Systems, Inc.
4700 Patrick Henry Dr.
Santa Clara, CA 95050

C

Calrad Imports
819 N. Highland Ave.
Hollywood, CA 90038

Cal-Tel Systems, Inc.
2674 S. Grand Ave.
Santa Ana, CA 92705

Pierre Cardin Electronique
1115 Broadway
New York, NY 10010

CCS Communication Control, Inc.
633 Third Ave.
New York, NY 10017

Celebrity Phone, Inc.
250 W. 57th St., Ste 1429
New York, NY 10107

Chrono-Art, Inc.
9175 Poplar Ave.
Cotati, CA 94928

Cobra Communications,
Div. Dynascan Corp.
6460 W. Cortland
Chicago, IL 60635

Code-A-Phone,
Ford Industries
Portland, OR 97228

Comvu Corp.
432 Park Ave. S.
New York, NY 10016

Contec Electronics, Inc.
150 N. State St., Ste 305
Chicago, IL 60610

Coreco Research Corp.
370 Seventh Ave., Ste 301
New York, NY 10001

Curley Corp.
915 Pennsylvania Blvd.
Feasterville, PA 19047

D

Dart & Kraft, Inc.
(See Anova Electronics)

Dictograph Corp.
4401 Waldon Ave.
Lancaster, NY 14086

Dur-O-Peg
933 E. Remington Rd.
Schauberg, IL 60195

Dynascan Corp.
(See Cobra Communications)

E

Electra Co.,
Div. Masco Corp of Indiana
300 E. County Line Rd.
Cumberland, IN 46229

Electro Audio Dynamics, Inc.
(See All Channel Products)

F

Fanon-Courier Corp.
15300 San Fernando Mission Blvd.
Mission Hills, CA 91345

"Firestik" Antenna Co.,
PAL International Corp.
2614 E. Adams
Phoenix, AZ 85034

The Fone Booth
12 E. 53rd St.
New York, NY 10022

Ford Industries
(See Code-A-Phone)

Fracom/Rovatone International
2130 W. Clybourn St.
Milwaukee, WI 53233

Arthur Fulmer, Inc.
Electronics Div.
122 Gayoso Box 177
Memphis, TN 38101

G

GC Electronics
400 S. Wyman St.
Rockford IL 61101

GCS Electronics, Inc.
7th Floor, Great Western Bank
Tower, 3200 Park Center Drive
Costa Mesa, CA 92626

Gemini Industries, Inc.
215 Entin Rd.
Clifton, NJ 07014

General Cable Co.
Cornish Div.
1701 Birchwood Ave.
Des Plains, IL 60018

General Communications &
Electronics Co.
59 Bloomfield Ave.
Pine Brook, NJ 07058

General Electric Co.
1285 Boston Ave.
Bridgeport, CT 06602

General Tele-Distributors,
30 W. Washington St.
Chicago, IL 60602

GTE
One Stamford Forum
Stamford, CN 06904

H

Hanabashiya Ltd.
39 W. 38th St.
New York, NY 10001

Harris Corp.
1680 University Ave.
Rochester, NY 14610

I

Integrated Circuits Packaging, Inc.
(Superphone)
750 N. Mary
Sunnyvale, CA 94086

International Dictating Equipment
Inc.
125 Wilbur Pl.
Bohemia, NY 11716

International Mobile Machines
Corp.
100 N. 20th St.
Philadelphia, PA 19103

Interquartz USA Ltd.
1700 N. Ashland Ave.
Chicago, IL 60622

Irvine Electro Sales Inc.
18207 E. McDurmott Drive
Irvine, CA 92713

Itera Ltd.
1535 Broad St.
N. Bellmore, NY 11710

ITT Telecommunications Corp.
133 Terminal Avenue
Clark, NJ 07066

J

Jasco Products Co., Inc.
217 N.E. 46th Box 466
Oklahoma City, OK 73101

Jerome Industries
8811 Shirley Ave.
Northridge, CA 91324

Jr. Enterprises
(Tel Time)
126 S. Illinois Ave.
Carbondale, IL 62901

K

Kendale Technology Corp.
4185 N.W. 77 Ave.
Miami, FL 33166

Keytronics, Inc.
786 Miraflores Ave.
San Pedro, CA 90731

Kir-Kat International Telephones, Inc.
855 Lexington Ave.
New York, NY 10211

L

Lake Communications, Inc.
5743 Howard St.
Niles, IL 60648

Lance Industries
13001 Bradley Ave.
Sylmar, CA 91342

M

Majestic Electronics, Inc.
14614 Lanark St.
Panorama City, CA 91402

Masco Corp. of Indiana
(See Electra Co.)

Maxon Electronics, Inc.
Airworld Center Complex
10727 Ambassador Dr.
Kansas City, MO 64153

MCE Inc.
23 N.W. 8th Ave.
Hallandale, FL. 33009

Megasonics Ltd.
39 W. 37th St.
New York, NY 10018

Metropolitan Teletronic Corp.
134 W. 18th St.
New York, NY 10011

Micro Communications Inc.
3307 Castor St.
Santa Ana, CA 92704

Midland International Corp.
1690 N. Topping
Kansas City, MO 64120

Mitel Corp.
Box 17170, Dulles Intl Airport
600 W. Service Rd.
Washington, DC 20041

Mura Corp.
177 Cantiague Rock Rd.
Westbury, NY 11590

N

Nady Systems, Inc.
Consumer Electronics Div.
1145 65th St.
Oakland, CA 94608

Paul Nelson Industries, Inc.
14 Inverness Dr. E. Bldg. 4
Englewood, CO 80122

New Horizons
1 Penn Plaza Ste 100
New York, NY 10119

Newcomm Electronics, Inc.
1805 Macon
North Kansas City, MO 64116

Nichco, Inc.
8660 Troy Township Rd 4, Rt 9
Mansfield, OH 44904

North Supply Co.
600 Industrial Pkwy.
Industrial Airport, KS 66031

Northern Telecom Inc.
Telephone Products Div.
640 Massman Dr.
Nashville, TN 37210

Nortronics Co., Inc.
8101 Tenth Ave. N.
Minneapolis, MN 55427

Novation, Inc.
18644 Oxnard St.
Tarzana, CA 91536

Nuvox Electronics, Corp.
150 Fifth Ave.
New York, NY 10011

O

Ohra Corporation
3555G Lomita Blvd.
Torrance, CA 90505

Olympia U.S.A.
Rt. 22, Box 22
Somerville, NJ 08876

Onyx Telecommunications Ltd.
505 8th Ave.
New York, NY 10018

P

Pacemark
Box 2223
New York, NY 10016

PAL International Corp.
(See "Firestik" Antenna Co.)

Panasonic,
Consumer Electronics Group
One Panasonic Way
Secaucus, NJ 07094

Pathcom, Inc.
24105 S. Frampton Ave.
Harbor City, CA 90710

Paul Nelson Industries, Inc.
14 Inverness Dr. Ste 6N
Englewood, CO 80112

Peace Electronics, Inc.
Box 873
Milpidas, CA 95025

Pfanstiehl,
3300 Washington St.
Waukegan, IL 60085

Phone Songs, Inc.
17 E. 96 St., Ste 14BV
New York, NY 10028

Phone-Mate, Inc.
325 Maple Ave.
Torrance, CA 90503

Phonesitter
10381 Jefferson Blvd.
Culver City, CA 90230

Phonies, Inc.
Box 2110
Cherry Hill, NJ 08003

Plantronics Santa Cruz
345 Encinal St.
Santa Cruz, CA 95060

PM Industries, Inc.
5946 Kester Ave.
Van Nuys, CA 91411

Precisa Products (USA) Ltd.
20610 Manhattan Place, Ste 108
Torrance, CA 90501

Proctor & Associates Co.
105050 N.E. 36th
Redmond, WA 98052

Q

Quintel, Inc.
1061 W. Van Buren St.
Chicago, IL 60607

R

Radio Shack
One Tandy Center
Ft. Worth, TX 76102

Record-A-Call
19200 S. Laurel Park Rd.
Compton, CA 90220

Recoton Corporation
46-23 Crane St.
Long Island City, NY 11101

Regency Electronics, Inc.
7707 Records St.
Indianapolis, IN 46226

S

Samhill Enterprises, Inc.
137 Fifth Ave.
New York, NY 10010

Sanyo Business Systems Corp.
51 Joseph St.
Moonachie, NJ 07074

Security Research Intl. Corp.
Intercenter 103, 160 SW 12th Ave.
Deerfield Beach, FL 33441

Sierra Electronics Corp.
50 S. Linden Ave. #6
South San Francisco, CA 94080

Skutch Electronics
209 Kenroy Lane #7
Roseville, CA 95678

Skyline Enterprises, Inc.
180 Northfield Ave.
Edison, NJ 08837

Sonic International Corp.
12 Greek Lane
Edison, NJ 08818

Sony Consumer Products
Sony Drive
Park Ridge, NJ 07656

Soundesign Corp.
34 Exchange Pl.
Jersey City, NJ 07302

SPS Industries, Inc.
John Hancock Center
875 N. Michigan Ave., Ste 3221
Chicago, IL 60611

STC Telecorp Inc.
(See Soundesign Corp.)

Strombert-Carlson
Charlottesville, VA 22906

T

Taylor Lock Co.
2034 W. Lippincott St.
Philadelphia, PA 19132

TDP Electronics
111 Old Bee Tree Rd.
Swannanoa, NC 28778

Technicom International Inc.
23 Gld Kings Highway
Darien, CN 06820

Technidyne Corp.
8550 Katy Freeway, Ste 216
Houston, TX 77024

Tectel, Inc.
14 Inverness Dr. E. Bldg. 4
Englewood, CO 80112

Tel Products Inc.
1513 13th St.
Lawrenceville, IL 62439

Tel Time
(See Jr. Enterprises)

Telco Products Corp.
44 Seacliff Ave.
Glen Cove, NY 11542

Tele-Com Products, Inc.
1058 N. Allen Ave.
Pasadena, CA 91104

Teleconcepts, Inc.
22 Culbro Dr.
West Hartford, CN 06110

Telecord 25
Box 698
Malibu, CA 90265

Telephone Extension Corp.
83 E. Central Ave.
Pearl River, NY 10965

Time Pen International Inc.
8700 Waukegan Rd., Ste 135
Morton Grove, IL 60053

Timely Products Corp./Telsec, U.S.
222 W. Adams St.
Chicago, IL 60606

T T Systems Corp.
9 E. 37th St.
New York, NY 10016

U

Uniden Corp. of America
6345 Castleway Ct.
Indianapolis, IN 46250

Unisonic Products Corp.
1115 Broadway
New York, NY 10010

Universal Security Instruments Inc.
10324 S. Delfield Rd.
Owings Mills, MD 21117

U.S. Tron
125 Wilbur Place
Bohemia, NY 11716

V

Valor Enterprises, Inc.
185 W. Hamilton St.
West Milton, OH 45383

W

Webcor Electronics, Inc.
28 S. Terminal Dr.
Plainview, NY 11803

Win-Tenna, Inc.
911 Amity Rd.
Anderson, SC 29621

Z

Zoom Telephonics, Inc.
207 South St.
Boston, MA 02111

INDEX